T0017570

Positively Green

Positively Green

Everyday Tips to Help the Planet and Calm Climate Anxiety

Sarah LaBrecque

National Trust

Published by National Trust Books

An imprint of HarperCollins Publishers, 1 London Bridge Street
London SE1 9GF www.harpercollins.co.uk

HarperCollins Publishers
Macken House, 39/40 Mayor Street Upper, Dublin 1, D01 C9W8, Ireland

First published 2023

© National Trust Books 2023

Text © Sarah LaBrecque 2023

Illustrations © Jem Venn 2023

ISBN 978-0-00-856763-7

10 9 8 7 6 5 4 3 2 1

All rights reserved. No part of this publication may be reproduced, stored in a retrieval system, or transmitted, in any form or by any means, electronic, mechanical, photocopying, recording or otherwise without the prior permission in writing of the publisher and copyright owners.

The contents of this publication are believed correct at the time of printing. Nevertheless, the publisher can accept no responsibility for errors or omissions, changes in the detail given or for any expense or loss thereby caused.

A catalogue record for this book is available from the British Library.

Printed and bound in the United Arab Emirates

If you would like to comment on any aspect of this book, please contact us at the above address or national.trust@harpercollins.co.uk

National Trust publications are available at National Trust shops or online at nationaltrustbooks.co.uk

FSC
www.fsc.org

MIX
Paper | Supporting responsible forestry
FSC™ C007454

This book is produced from independently certified FSC™ paper to ensure responsible forest management.

For more information visit: www.harpercollins.co.uk/green

Contents

Introduction

If you're like me, your relationship with 'green living' will be complicated. Those feelings of impassioned motivation to be as sustainable as possible can, at times, take a back seat to the demands and distractions of everyday life.

I have a young daughter and work as an editor for a magazine. I have washing up to do, commitments with family and friends to keep, and toys that need to be tidied away (again!).

But bleak headlines about the climate crisis and the state of nature keep rolling in – serving as reminders on an almost daily basis that things aren't looking good. As a society, we're going to need to make some changes. And I know I'm a part of that.

Luckily, despite sometimes feeling that I should be doing more, there are certain eco habits that have been so deeply ingrained in me that I know I'll never lose sight of them. My parents instilled in me a deep sense of 'waste not, want not'. Leftover food was carefully packed away and broken toys were patched up; dresses worn by my older sister – themselves passed down from family friends with older daughters – would eventually find their way to me. Despite this early fostering of environmental stewardship, I was, and inescapably remain, a part of the global consumption machine. As we all do.

As the climate crisis ramps up, it's easy to get caught down frightening rabbit holes – the world's precipitously decreasing levels of biodiversity, for example, the microplastics in our waterways, or the mountain of e-waste accumulating in the global south.

But there is reason for optimism: about what can be done if we, as individuals, take personal responsibility for the things in our control. Incredible power for change lies in individual and household behaviour. According to one European study (funnily enough, based on data from something called the HOPE project) household consumption was found to account for 72 per cent of global greenhouse gases (GHGs).

Consider that for a moment.

The way *you* shop, dispose of waste, use energy and eat has a direct bearing on the globe's future ability to regulate its climate. Of course, businesses and government have a huge responsibility in this too – they provide your household's goods and services, after all – but if we're talking about what individuals can take ownership of, it's significant.

But don't let that burden you. Instead, feel empowered. History has shown that when groups of people unite together under a common goal, particularly when there's a lot at stake, paradigm-shifting change is possible.

Take the Montreal Protocol, for example. In 1987, all 197 UN member states signed this binding treaty – the first ever to be adopted by every country in the world. It would set into motion the mitigation and eventual elimination of harmful chemicals that were destroying the protective ozone layer and is widely considered to be one of the most successful environmental treaties – ever.

Such multilateral, high-level action sounds far removed from an ordinary person's everyday reality. But, in essence, what happened was that a group of people recognised the seriousness of a problem. And did something about it.

Examples of concerted action resulting in meaningful change are everywhere.

Rewilding efforts by conservationists in the UK are slowly bringing wild beavers back from the brink; the UK's plastic bag tax has translated into an unprecedented 97 per cent reduction in the single-use variety; and during the Covid-19 pandemic the global community rallied and supported one another in a way many of us had never witnessed before.

Meanwhile, people-powered movements across the country are ushering in a new way of living within our environmental means and harmoniously alongside nature. The pandemic saw an army of litter-pickers band together to clean up the countryside and waterways, and there are now hundreds of 'zero waste' stores across the country. Back in 2017 there was one.

There are heartening examples aplenty that point towards people's desire to live green. Nevertheless, there's still so much to be done – and the climate crisis will wait for no one.

With the help of this guide, I hope that instead of feeling anxious about the fate of the planet, or sceptical about how any one individual can make a difference, you'll be gripped by a sense of highly personal empowerment which gives rise to a steadied calm.

'We aren't necessarily insignificant,' says Professor Tim Lenton, who is studying 'positive tipping points' at the University of Exeter. Usually discussed in reference to the greatest threats humanity faces – ice sheet disintegration, Amazon rainforest dieback or coral reef loss – tipping points happen when small changes combine to cross a critical threshold, triggering irreversible consequences. But Lenton and his team argue that the same principle is already providing solutions and positive change. In Norway, an electric vehicle (EV) tipping point has been reached: they're now cheaper than petrol cars and more than half of vehicles bought there are electric. In the UK, power generation is decarbonising faster than in any other large country, with positive tipping points playing a role.

'This concept could unlock the stalemate – the sense that there's nothing we can do about climate change,' says Lenton. What does this mean? Your incremental lifestyle changes matter, because they're part of a process that's bigger than you. All you need to do is embrace that.

VINEGAR

A Green Home

For many people, life has become a lot more home-based over the last few years. The pandemic confined a large proportion of us to our domestic abodes, but spending all that time at home gave us a chance to consider how best to run them. Or green them.

It's useful to ask yourself: what's coming in? And what's going out? If you're using harsh cleaning products, some of them are likely being released back into the environment when you slosh wastewater down the sink or toss soiled kitchen towels in the bin. If you're powering your household with non-renewable fossil fuels, what's coming 'out' – in a manner of speaking – is climate change-inducing emissions.

The analogy can apply to almost anything. Buy in lots of hyper-convenient but heavily-packaged goods by post and the odds are that most of that packaging will end up sitting on your kerb the next week. A good rule of thumb, when considering what you're bringing into and sending out of the home, is to aim for ***none***, ***natural*** or ***neutral***.

That is, first consider if you need the item or service in the first place. If you can't eliminate it completely, try using less of it or go for natural options. As a third option, choose products that have sustainability credentials: things that are recyclable, compostable, made from repurposed materials or come from a renewables-powered factory, for example. I like to think that such attributes help to *neutralise* a product's impact. ***None***, ***natural*** or ***neutral*** – stick that on your fridge.

Going further, if you can find *carbon neutral* products, excellent. But that's still pretty challenging. And it's difficult to work out whether companies are as green as they claim to be. Nonetheless, a growing group of businesses and organisations, small and large, are committing to net zero emissions (a related concept). This, and being 'net positive', is the next frontier of sustainability.

A quick word on net zero

In order to reach the level of reductions required to avoid irreversible climate change, every part of society, every business and every household will, in time, need to be 'net zero'. Strictly speaking, the term relates specifically to emissions: it means that they have been reduced as closely as possible to zero, and that any left behind are removed from the atmosphere using carbon offsetting. But we can also apply net zero thinking to life as a whole, so that our impact on the environment and climate is essentially nil.

An even more ambitious goal is to be 'net positive', where people give back more to society and the planet than they are taking away. For instance, through regenerative agriculture; houses that create a surplus of renewable electricity through on-site generation; or a business that contributes more value to its community and the environment than if it otherwise wasn't there.

Despite the fact that the concept has been discussed in corporate sustainability circles for over a decade, the reality of achieving this kind of net positive society is some way off.

In the meantime, when it comes to greening your household, stick to ***none***, ***natural*** or ***neutral***.

And as far as (net) positivity goes, let that be your north star and infuse your mindset.

General cleaning

Over the course of the pandemic, people went cleaning crazy. Never a fan of powerful disinfectant cleaners myself, during those first few months of Covid, even I was swept up in the frenzy. If the bottle wasn't stamped with the authoritative 'Kills 99.9 per cent of viruses and bacteria!', were we knowingly inviting Covid-19 into our homes? I remember going to the supermarket one day, the shelves stripped of pasta and loo roll, not a bottle of disinfectant left in sight.

Thankfully, the panic has subsided and, according to market research, the longer-term trend among the public for greener cleaning products is again back on course. By 2026, the global sustainable cleaning market is forecast to be worth nearly $110bn.

None

For general household cleaning – kitchen surfaces, dusty bedside tables and the like – sometimes a simple damp cloth works just fine. Before grabbing the spray bottle, ask yourself if you really need de-greasing or de-scumming action. If it's just surface dirt to be wiped and swiped, perhaps plain and simple water will do.

Natural

There are plenty of DIY and natural solutions out there, with bicarbonate of soda, vinegar and castile soap as three of the products that should be go-tos for any green home cleaner.

Distilled white vinegar Diluted with water this is an effective, non-toxic and biodegradable cleaner that is versatile to boot. If you're not a fan of the smell, you can add a few drops of essential oil – but be aware that some are toxic to pets, so do avoid essential oils if you have furry friends at home.

Bicarbonate of soda This can be used to scrub toilets, soften caked-on food on pans and in ovens, and clear drains. The key to bicarb is to leave it on for a while to work its magic, then scrub with water.

Castile soap Another incredibly versatile, natural, vegan and biodegradable alternative to conventional cleaning products. Made from vegetable oils such as hemp, avocado, coconut or olive oil, castile soap originated in Spain and comes in liquid or bar form. Despite avocado and coconut oil tending to have a higher environmental footprint than other plant oils, castile soap is thought to be an all-around more natural and sustainable choice than conventional detergents. It can be used for all sorts of jobs: mopping floors, cleaning dishes, shampooing pets and rinsing fruit and vegetables. You can even use it as a hair or body wash. You'll usually need to dilute with water so do check online or read the pack instructions for directions.

More natural cleaning ideas

Here are some other simple, all-natural recipes.

Drain cleaner

The key to keeping drains clean and clear is to take a preventative approach and unclog them regularly. Sprinkle half a cup of bicarbonate of soda down the drain, followed by 250ml of white vinegar. Leave to fizz for 10 minutes and then pour 4 cups of boiling water down the drain.

Oven cleaner

Simply sprinkle some bicarb over caked-on food stains and spray some water on top. Leave for an hour and then scrub off with a scouring pad. Leftover pieces of aluminium foil also make handy scrubbers.

Hand soap

Combine 500ml of water with 1–2 tablespoons of liquid castile soap, a few drops of essential oil (optional) and pop in a soap dispenser. Using a foaming hand-soap dispenser will make your soap go further, and if you find the formula too thin you can add a saltwater solution to thicken it up.

Dishwasher liquid

Combine 125ml of liquid castile soap with 125ml of water. Then add 1 teaspoon of lemon juice, 3 drops of tea tree extract and 60ml of white vinegar. Use 2 tablespoons per load. (Courtesy of Friends of the Earth UK.)

Multi-purpose cleaner

Put the rind of lemon (or any other citrus fruit), with the pith removed, in a clean glass jar. Add a woody herb, such as rosemary or lavender. Top with white vinegar and infuse for two weeks. Strain through a cloth into a spray bottle and dilute with water to make a 50:50 mix.

(Courtesy of Kate Jones' Instagram account, @my_plastic_free_home – a great resource for natural and DIY household tips.)

Neutral

Greener cleaning products are fairly easy to find these days, with many companies swapping harsher chemicals for more nature-based ingredients. Try to choose products that are free from phosphate, phosphonate and other strong chemicals such as chlorine, and aren't tested on animals. Look out for those stamped with the EU Ecolabel or an International Association for Soaps, Detergents and Maintenance Products (AISE) 'Cleanright' sustainability mark. Bottles made from recycled plastic and those that can be refilled are two more neutralising wins.

Recycled plastic will increasingly be included within plastic packaging as big corporates make progress on their sustainability targets and respond to government legislation. Companies that produce or import plastic packaging that does not contain a minimum of 30 per cent recycled content are subject to a government tax that was introduced in April 2022.

But as recycled content goes up, the cleaning aisle might start to look a tad more drab. Have you ever noticed how the packaging of eco cleaning products tends to be less colourful? This is because the bottles often contain a higher proportion of recycled plastic, which is currently more challenging to produce in a vibrant palette. For the big brands, consumers' unconscious bias towards pretty packaging is a serious hurdle in terms of marketing more sustainable but less attractive alternatives, as silly as that seems.

With that in mind, be aware of your own biases. When effective cleaning action is all you require, who cares what

Cleaning with probiotics

In light of growing concern around antibiotic resistance and increased awareness of the importance of a healthy gut for overall health, research around probiotics (as well as pre- and post-biotics) is burgeoning. So too are the number of consumer products boasting billions of friendly bacteria ready to colonise everything from your skin to your stomach. For the inside scoop on what goes on on the inside, I highly recommend Giulia Enders' book *Gut*. It's a fascinating account of 'our body's most under-rated organ' with delightfully enlightening chapters such as 'A Few Facts About Faeces' and 'I am an Ecosystem'. There's also a section on the relationship between good bacteria and cleaning.

Indeed, it's not just supplements and skin creams that promise to fine-tune the balance of bacteria (albeit with varying degrees of robust evidence to support their efficacy, so do be wary of claims). Scientific research at the University of Ferrara in Italy found that when probiotics were used as a cleaning agent within a hospital, not only were they effective but they continued to 'clean' long after being applied.

It works like this: with every spray of a probiotics-infused cleaner, billions of 'good' bacteria are released. But rather than descending on their enemies – 'bad' bacteria and viruses – they instead eat the baddies' food and home, feasting until there's nothing left for the unwanted bacteria and viruses to live on, or consume. Probiotics effectively outcompete the undesirables, staying active on surfaces for potentially weeks at a time. Conversely, levels of microbial contamination can return to pre-cleaning levels within 30–60 minutes of applying conventional cleaners, according to Elisabetta Caselli, the author of the Italian study.

It's a relatively new area of research but definitely one for green-cleaning enthusiasts to watch. There are now even some UK-based companies that offer a range of probiotic cleaners via subscription and in reusable bottles.

the cleaning products look like? After all, they sit in a dark cupboard and not on your mantelpiece.

tip:
Old, cut-up t-shirts, socks or sheets make good general cleaning rags.

Of course, it's not just about cleaning products. It's also about what you use to apply them. These days, disposable kitchen towels and blue cloths are out, and biodegradable, reusable cloths and sponges are in. Now easy to find at any zero waste store or online, these highly absorbent alternatives are typically made from a mix of natural fibres and can be tossed into the washing machine when dirty.

Laundry

Washing. It's endless, it's essential, and it comes with a sizeable environmental price tag. From plastic microfibres escaping from our clothes to water and detergent use, laundry's impacts stack up. But for a machine that UK households use on average once every one and a half days, it's surprising to note that most of its carbon footprint doesn't actually come from use but rather from manufacture and delivery – up to 80 per cent. It's therefore safe to say that the most eco-friendly machine is the one you already have. Don't splash out on a new one until yours is well and truly beyond repair.

Luckily, new 'right to repair' legislation that came into force in 2021 in the UK means that manufacturers of white goods, such as washing machines, dishwashers and washer-dryers, must make available spare parts to consumers and third-party companies. Therefore, it's always worth exploring repair as a first port of call.

If your machine is beyond help, splashing out on a higher-quality one could save you money in the long run. They are generally more energy efficient and therefore cheaper to operate. Opt for those with A or B energy ratings and try to launder at lower temperatures. Washing with colder water can be effective for lightly soiled clothing, and all machines should have at least a 20°C temperature setting by law. Helpfully, detergents these days are formulated to work at lower temperatures. Plus, for colourful clothing or darks, cold water washes mean dyes are less likely to run.

A few more washing wins

- Avoid doing half loads.
- Take care of your machine: clean the filter monthly and don't overload the drum.
- Wash your clothes less often.
- Don't tumble dry unless you really need to.
- Check out laundry eggs or balls. These alternatives to detergent come filled with mineral pellets that act as a natural cleaning agent.

Microfibres

In October 2014, *The Guardian* published an article entitled 'Inside the lonely fight against the biggest environmental problem you've never heard of'. It became one of the most widely read articles of the year for the sustainable business section of the newspaper (something I happen to know as I was a member of the team who worked there at the time).

Nowadays, most people have heard of microfibre pollution – and hopefully those working tirelessly to combat it are feeling less lonely. But it's still a serious problem.

The tiny fibres that shed from fabrics during washing, and particularly tumble drying, are microscopic menaces to the environment; they have been detected in food, water and even in Antarctic sea ice and very high up in the

Earth's troposphere. Research around the true prevalence and effects of microfibre release are ongoing: scientists are looking at how potentially pathogenic bacteria are attracted to microfibres, for instance, and what effect this might have on human and animal health. As pollution around coastal areas and ocean temperatures rise, bacterial growth could be exacerbated.

Meanwhile, work is afoot to urge manufacturers to kit out machines with filters that can catch microfibres. And independent startups are ploughing ahead with the development of microfibre-catching gadgets that could be fitted to existing washing machines.

But while that's in progress, there are things you can do.

Managing microfibre pollution

- Follow the 'washing wins' tips opposite.
- Invest in a washing bag that collects microfibre particles in your washing machine, to be scooped out and put in the bin (slightly better than being flushed out into wastewater, but not a perfect solution).
- Choose organic or recycled natural fibres over synthetic fabrics if you can.
- Avoid microfibre cleaning cloths.

Positivity Pause

Climate anxiety is a rising concern, especially among younger generations. In 2021, *The Lancet* published a far-reaching study that explored the extent of eco-anxiety, and its relationship with government action. Of the 10,000 16–25-year-olds surveyed, three-quarters said the future was frightening. Which, in itself, is frightening.

Britt Wray, who worked on the study and is the author of *Generation Dread*, a book about finding purpose in the age of climate crisis, talks about how working through her own psychological turmoil around the fate of the planet led her to pursue 'more nourishing and radically helpful narratives about the future'. Alongside her book, she also publishes a newsletter called *Gen Dread* that she describes as 'the clearing house for new, old, and emerging ideas to strengthen our emotional intelligence, psychological resilience, and mental health while we're in this planetary pickle.' If you're experiencing eco-anxiety, it's well worth checking out.

You could also consider the wisdom of writer Oliver Burkeman, set out in his book *Four Thousand Weeks* (roughly the number of weeks the average person has on this planet). Aimed primarily at those who grapple with the nagging need to be ever more productive, some of his thinking can be applied to quell eco-anxiety.

Burkeman espouses the idea that we cannot do more than we can do. By understanding our limitations and accepting that there will always be more – more emails to answer, more health hacks to adopt and green living tips to try – we can liberate ourselves from the unhappy hamster wheel of continually feeling disappointed about what we haven't achieved.

Few of us will be able to become the zero waste warriors we read about – the family whose accumulated rubbish over the course of a year was only enough to fill a jam jar. Or the couple who grow all their own food on their renewables-powered, off-grid homestead. But there are myriad habits we can relatively easily incorporate into our lives. One by one, bit by bit.

Try concentrating on one of the sections or chapters in this book per month. Or at intervals that work for you. Or perhaps focus on the areas that will have the most impact, which are food, transport and heating.

As Burkeman says, '... the only route to psychological freedom is to let go of the limit-denying fantasy [that you'll be able to get] it all done ... instead focus on doing a few things that count.'

Waste and recycling

Of all habits done in the name of environmental responsibility, recycling is probably the most well-embedded. According to a 2021 survey by the Waste and Resources Action Programme (WRAP), 88 per cent of households said they 'regularly recycle', and more than half reported that they recycle more than they did a year ago. Nevertheless, England isn't a top recycler on the European stage. It only recycled 44 per cent of its waste in 2020, down from 46 per cent the year before. But three cheers if you live in Wales: it achieved a 56.5 per cent rate, the only UK nation to exceed the 50 per cent target set by the European Union for 2020.

But while recycling is a tried, tested and well-understood way to recover the value from resources and prevent waste to landfill, it gets arguably too much airtime in relation to something that's far more all-encompassing: the circular economy. A traditional linear economy is defined by a 'take, make, waste' trajectory, but a circular economy aims to keep resources in circulation for as long as possible, while doing away with waste and regenerating nature. It very much incorporates ***none***, ***natural*** or ***neutral*** thinking. But what exactly does it mean in practice?

A circular-economy-centred approach can be applied to everything from the construction industry to chemistry, but in relation to household waste, it's really about the hierarchy of things. Namely, that prevention is the best waste strategy there is, and efforts simply to stop using what you don't need or reducing waste-generating items should be at the top of your priority list. Failing that, we should prioritise buying items made from recovered materials, and reuse, reuse, reuse.

Easy swaps for a more circular kitchen

- Cover bowls of leftovers with a plate or beeswax wrap instead of clingfilm.
- Use reusable (and eventually biodegradable) kitchen cloths and towels.
- Use casserole dishes with lids rather than cooking food in aluminium foil.

A note on food waste

There are oodles of creative ways to reduce food waste and use up leftovers, which we'll delve into in a later chapter. But if you're not able to set up a home compost (skip to page 114 for composting tips) using your council's food waste collection is an absolute no-brainer.

As part of the government's target to 'eliminate all avoidable waste' by 2050, every household in the UK should have separate, weekly food waste collections in the next few years.

Globally, about a third of all food produced goes to waste – the emissions associated with the production, transport and gases released when it rots account for 8–10 per cent of the world's GHGs. Being resourceful at home, and only buying and cooking what you need, is key.

Bathroom and beauty

Whether it's waterless shampoo bars or buying products with nature-based or organic ingredients, there are lots of opportunities to make your bathroom a little greener. There is a sliding scale of commitment to the cause, with some very dedicated DIYers concocting all sorts of homemade recipes for personal care products and buying the best in natural makeup and the like. If you have the time and inclination, there are online tutelage options available where beginners can work towards certificates in formulating natural and organic skincare products. But for most of us, who just want some quick tips for eco-fying our beauty routine, here's a checklist:

Solids

Solid shampoo, conditioner and soap bars are exploding in popularity. Found in zero waste stores and from retailers such as Lush, who pioneered the concept, using solid products means fewer transport emissions due to lighter shipping (shampoo, for example, is up to 80 per cent water), and reduced or no plastic packaging. Be conscious of lingering longer in the shower while you lather up though. Try turning off the water after you've got a good foam, then on again to rinse.

Refillables

Makeup, deodorant, haircare products and more can now be found in the refillable variety. While an excellent idea on paper, do consider carefully which refillable products make sense for you. Experts estimate that you need to refill a product between

50 and 100 times before its impact justifies the extra packaging often required (refillable packaging is more durable than disposable packaging). So, if you're certain of your loyalty to a particular brand – you know you'll always buy your lipstick from Lush, for example, who sell a refillable cartridge, then by all means, refill away.

Reusables

The same principle of being sure before you invest applies to reusables too (admittedly difficult if you've never tried something before, but most of us have a general sense of whether something is right for us or not). Reusable period products in the form of menstrual cups or super absorbent 'period pants' could save you thousands of pounds over the course of your lifetime, and collectively about 200,000 tonnes of waste to landfill, too. Washable makeup-remover pads are another good idea. And for babies, it's worth investigating non-disposable wipes and nappies. Ditching disposable razors is another easy habit swap.

Recycle

For used packaging, it goes without saying that you should recycle anything you can. It's also worth investigating what recycling drop-off schemes are available in your area for tricky-to-recycle items such as makeup packaging.

32

Energy

According to the Energy Saving Trust, over a fifth of the UK's carbon emissions come from homes: how we heat and cool them, power up our devices and appliances, and generate hot water. But there are significant targets in place to cut this down to nil, in time.

By 2035, the entire electricity network will be powered by green energy, or so the government intends. And by 2050, heating-related emissions, which account for over half of overall household emissions, should be reduced by more than 95 per cent.

It's going to require a huge overhaul of infrastructure and technology – of the gas network and the way boilers work, for instance. Not to mention a total shift away from fossil fuels – no small task.

It will be fascinating to watch all of this play out in the next few decades, but, in the meantime, there are lots of ways you can take control over your own energy footprint, for maximum reduction and efficiency. You may even save a pound or two.

Keep the heat in and cold out

When it comes to making your home as energy leak-proof as possible, bigger ticket investments such as insulating, draught-proofing and buying an energy-efficient heating system can go a long way. Based on a typical mid-terrace house, as calculated by the Energy Saving Trust, you could make the following savings by adding insulation:

Insulate	Cost savings per year
Walls	£180–245
Loft	£230
Tanks, pipes, radiators	£160
Total:	£570–635

(Energy savings estimates based on April 2022 calculations.)

There are also many smaller energy-wise actions that, added together, equate to hundreds of pounds saved. The Energy Saving Trust has also calculated that you could save roughly £20 a year by switching off lights, £36 by not overfilling your kettle and £70 by taking four-minute showers.

Turning your thermostat down by just 1°C is another easy and painless way to reduce energy and bills. You could save up to £128 a year and all you need to do is put on another layer.

Smart meters

The UK government originally set a target for all households to have smart meters installed by 2020, but the rollout programme has been beset by delays. Gas and electricity suppliers are now required to have kitted out all remaining non-smart-metered customers by 2025. At only five years behind schedule, when the target is reached it will be momentous – by all means, when the installation expert comes a-knocking, do raise a glass.

Evidence suggests that once people have a smart meter, they are more likely to make energy-saving changes: a survey by Smart Energy GB found that this was true for 85 per cent of households with the energy tracking devices. In the future, when we're more reliant on a renewables-powered grid and higher numbers of homes self-generate green power, smart meters will form an important part of a digital network that keeps tabs on energy ebbs and flows.

If you've got one installed already and are keen to dive deeper into how it might help you become more energy savvy, try the Loop app. This free tool connects to your smart meter data and helps you to see how much energy you're using across different time periods. It's also got tips for reducing 'phantom load' – the amount of power that's used when appliances or gadgets are switched off or on standby mode – as well as a solar simulator. By entering some simple details about your home and roof, it can tell you how much you might save if you had solar panels.

Renewables

Which leads us to renewables, our fine, fair-weathered friends. If all goes as planned, all our electricity will come from green sources by 2035 – but there's an awful lot of work to be done to make that goal a reality, from the quadrupling of offshore wind capacity to the large-scale rollout of battery storage.

From an individual's perspective, it may seem like there's not much influence you can have on industrial-scale plans to fully decarbonise the grid. But there are meaningful actions you can take.

Install green energy systems in your home

From solar panels to hydrogen boilers, domestic renewables technologies for heating and electricity come with varying price tags (and degrees to which they're available).

The most common type of on-site household-level generation, by far, is solar. About 1 million UK homes now have panels installed, which is just over 3 per cent of all homes across the country. Demand has surged in recent years, bringing prices down, but outlay costs are still relatively hefty. The average solar installation will cost about £6,500 but if you sell excess electricity back to the grid, savings could be in the range of £300–£500 per year.

For green heating systems, there are a range of options, from heat pumps to solar hot water heaters, but, again, upfront costs are fairly high. There is some financial support available from the government, but it's not extensive. Grants of £5,000 for heat pumps, for example, are available on a first-come, first-served basis until 2025, to help cover installation costs, but you'll need to do your homework around whether they're suitable for your home, and whether it makes financial sense to invest. Do seek advice and information from a reputable source such as the Energy Saving Trust.

Invest in green power

A slightly less involved but no less valuable way to support the renewables transition is to invest in organisations that are facilitating it. Bristol Energy Cooperative (BEC) is one such group. It provides organisations with free solar panels and supports renewables projects across the city. It periodically runs community share offers, which are open to anyone and typically require just £100 as a minimum investment. If you're based in England, check out communityenergyengland.org for share offers that are open with similar organisations across the country. Community Energy Wales, Community Energy Scotland and Northern Ireland Community Energy also have relevant information.

Some investment platforms enable people to put their money towards a range of climate-friendly initiatives, from EV charging infrastructure companies to solar, wind and energy storage projects. Annual returns are in the 2–10 per cent range, depending on the project, but always make sure you do your research before parting with any hard-earned cash.

Go on a renewables tariff

With energy prices at record highs, it's understandable that most people are prioritising cost over anything else. It may seem counterintuitive, but renewables tariffs are generally subject to the same price hikes as standard ones, because all energy companies, regardless of what type of tariff they offer, are at the mercy of fluctuations in the wholesale market. So, renewables tariffs are unlikely to be cheaper than standard ones – and may even be more – a slightly painful truth that makes the argument for switching trickier.

But at the end of the day, choosing a truly green energy supplier that directly supports renewable generators (search *Which?* for their list of 'eco-provider' energy companies) and the transition as a whole is an incredibly impactful thing to do. Why?

There are positive feedback loops currently at play that mean renewables are growing exponentially on a global scale. As demand rises, more capacity is deployed, and prices come down. The more attractive the economics, the more investment there is, and so on and so on. Seeing yourself as part of these positive feedback loops – or positive tipping points, à la Professor Tim Lenton and his research on such things – is a thoughtful way to consider the impact of going with a renewables tariff, or, indeed, installing a green power system in your own home. You might not immediately see a benefit to your purse, but your contribution is part of a powerful onwards force that looks to be unstoppable. And money savings will eventually follow.

In 2012, the International Energy Agency predicted that worldwide solar energy generation would reach 550 terawatt-hours by 2030. But that figure was exceeded 12 years earlier than anticipated. Meanwhile, there are social factors at play that could be contributing to positive feedback loops and future tipping points: researchers in the USA have found that installing solar panels is contagious. Studies demonstrated that when people had the panels put in, the odds increased that neighbours would follow suit. In the city of Fort Collins, Colorado, 69 per cent of installations came from referrals.

This kind of keeping-up-with-the-Joneses phenomenon is consistent with Lenton's work around positive tipping points. 'Social contagion underlies accelerating adoption of some sustainable behaviours – for example, the uptake of rooftop solar photovoltaic (PV) systems and other household energy behaviours,' he says in his exhaustive study. In other words, the sum of collective action on green energy is bigger than its parts – so try not to think of your household in isolation.

But do be diligent when it comes to picking a renewable energy provider. Greenwashing, which is essentially when companies peddle misinformation around their sustainability credentials, is rife in the sector. The regulator Ofgem is currently doing an investigation into renewable electricity deals, and whether those that claim to be 'green' or '100 per cent renewable' are in fact so. Uswitch's Green Accreditation site is a good place to do some comparison shopping.

You can also switch your energy supplier if you live in rented accommodation. As long as you pay your gas and electricity, it's within your rights to change supplier.

Links

energysavingtrust.org.uk/a-quick-guide-to-low-carbon-heating

energysavingtrust.org.uk/energy-at-home/generating-renewable-electricity

energysavingtrust.org.uk/advice/solar-panels

uswitch.com/gas-electricity/green-energy/green-accreditation

More on money

When it comes to green actions, the conversation tends to centre around lifestyle changes, but don't overlook the power of your pennies to effect change.

Here are three ways to harness the green power of money:

1. Green your pension

According to the Make My Money Matter campaign, for every £10 put in pensions, £2 is linked to deforestation. Other harmful industries such as arms, tobacco, fossil fuels and gambling benefit from the £2.7 trillion being poured into UK pensions. Choose a provider that offers fossil fuel-free and socially responsible options – Ethical Consumer can recommend some. And if you're already enrolled in a pension via your workplace, urge your pension provider to ditch investments in planet-damaging sectors by signing the Make My Money Matter petition.

2. Choose a responsible insurance company

The insurance industry has disproportionate power over the future of the planet, through their investments and underwriting activity. Some of the major firms have ceased to insure new fossil fuel projects but many still invest in them. When you're seeking out insurance, for your home, travel or car, for instance, go with a provider that takes a green approach. Lori Campbell's blog good-with-money.com offers some useful recommendations.

3. Seek out an ethical bank

High street banks aren't known for being environmentally or socially responsible. In fact, some of the country's most popular banks finance fossil fuels and have been linked to human rights abuses. If you're interested in reading more on this, check out the annual Banking on Climate Chaos report.

Luckily, there are some very reputable and ethical alternatives to choose from. The rankings and research body Ethical Consumer makes it easier to find banks which, for instance, finance sectors such as renewables, sustainable farming and charities – and are transparent about who they lend to.

While switching accounts might sound like a faff, using the Current Account Switch Service actually makes the process fairly seamless.

Links

ethicalconsumer.org

makemymoneymatter.co.uk

currentaccountswitch.co.uk

Food and Drink

'Swapping just one meat dish for a plant-based one saves greenhouse gas emissions that are equivalent to the energy used to charge your phone for two years. Your small change can make a big difference.'

This was a message that was printed on menus as part of a study involving 6,000 people in the USA, conducted by the World Resources Institute. And it's attention-grabbing for several reasons.

First, it bears pause for thought at how very unbelievable – and exciting (at least for eco-nerds like me) – the statement is. One meal. Two *years* of phone charging.

Now that you've taken 10 seconds to run that through your mind, let me tell you what happened when diners read it. They chose a vegetarian dish 25 per cent of the time, which is more than double the rate of patrons who were shown no message. It's an argument for the power of clever framing to nudge people in the direction of climate-friendly meals. People are now accustomed to seeing this kind of messaging on products and, argued the researchers, are primed and ready to make changes. But it helps enormously if nudges are especially empowering or appetising.

Think about how veggie options are usually presented on menus: 'vegan', 'healthy', 'meat-free'. No, my mouth's not watering either. Alongside messaging that shows how small changes can have a big impact, or how a growing group of people are joining a movement, using more descriptive adjectives could also make a huge difference in mitigating the cumulative climate effect of people's food choices. The food industry accounts for more than a third of global emissions after all, with meat responsible for over half the planet-heating pollution of grains, fruit and veg – so we're not talking small potatoes here.

You're likely not in the business of writing menus yourself, but simply being aware of your own bias – against meals that sound less than appetising, which, admittedly, no one can fault you for – could help you to see things differently. The 'healthy vegan bean burger' might not sound like anything to write home about (and indeed, the author of the menu on which said burger appears needs some schooling), but it may actually be juicy, flavourful and satisfying. You won't know until you try.

But I'm not here to convince you to become a vegan – eating habits are deeply ingrained, highly emotive and influenced by complex cultural and socio-economic factors. And while meat and dairy consumption contribute disproportionately to diet-related emissions (it is important to come to terms with that, mind) it is possible to adopt a more flexible approach and still make a significant impact.

Flexitarianism, which is, roughly, being open to all sorts of diets – eating plant-based meals one day and tucking into a chicken burger the next – has grown in popularity in recent years. And climatarianism, where you eat according to food's climate impact, is another eco-friendly way to consume. Climatarians might choose to eat locally produced meat, for example, but eschew almonds because their production typically requires very high water use.

Whatever diet you choose, one guiding principle is clear: eat more plants. Research suggests that doing so, as well as following standard dietary guidelines – more fruit and veg, wholegrains, lean protein and legumes – could result in a 29–70 per cent reduction in food-related GHG emissions. So, throw some (British-grown) strawberries onto that muesli.

Local, organic, seasonal and beyond

When people start talking about the organic veg box they receive from their local community-supported agriculture scheme, replete with biodynamic wine and rare, heirloom tomatoes, it can all start to sound disappointingly ... middle class. The arguments for eating organic, with the seasons, and from nearby farms are all sound: biodiversity-harming pesticides are left behind, food is less intensively produced and food miles and their associated emissions are reduced, respectively. Buying from local farmers also supports a more resilient economic system that is potentially more protected from price shocks. And food that hasn't travelled as far tends to taste better too.

But will eating sustainably leave a hole in your wallet? Not if you're strategic about it, no.

Top tips for budget- *and* planet-friendly eating

Buy wonky

Do your bit for curly courgettes, poky peppers and piddly potatoes by scooping them up and giving them a happy end (your stomach, rather than landfill). Most major supermarkets now have misshapen ranges that are typically cheaper than their more conventionally attractive counterparts. If you're doing your shopping online and are struggling to find wonky fruit and veg, it might help to filter results from lowest price to high. Subscription boxes such as Oddbox also deliver wonky veg straight to your door.

It may not be organic, or sourced from a farm around the corner, but reducing food waste is a big part of mitigating your climate footprint. According to WRAP (Waste and Resources Action Programme), if food waste was a country, it would have the third highest carbon footprint after the USA and China. So arguably, reducing food waste is just as, if not more, important than being selective about what you eat.

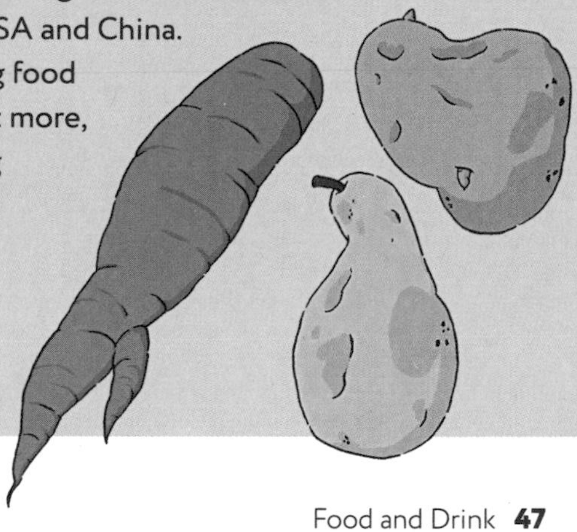

Shop seasonally – at the supermarket

Many of us are creatures of habit when it comes to what we put in our stomachs, and children are often picky. But buying and cooking the same meals all year round doesn't always make sustainability sense. Veg boxes usually feature fresh, seasonal produce, but don't overlook supermarkets for good prices on in-season, British-grown fruit and veg. By occasionally doing your shopping in-store as opposed to online, you'll also better be able to see what's cheap and abundant for the time of year. Try simple swaps in favourite dishes according to the seasons: spring greens in your stir-fry in April and May, for example, and kale in November.

Audit for freshness regularly

No matter if it's locally produced, seasonally grown or organically cultivated: if it's going blue in your fridge, it's a problem. Combat unnecessary waste by regularly checking best before and use-by dates and shuffle your weekly meal plan accordingly. You'll save money by not having to toss as much away.

Grow or make your own

Cooking more from scratch can be another way to save money and resources (although not necessarily time!). You can make cheap and cheerful meals from dried or tinned pulses, and baking your own bread will work out more affordable than shop-bought. Growing fruit and veg from seed is also very economical and comes with no sustainability strings attached.

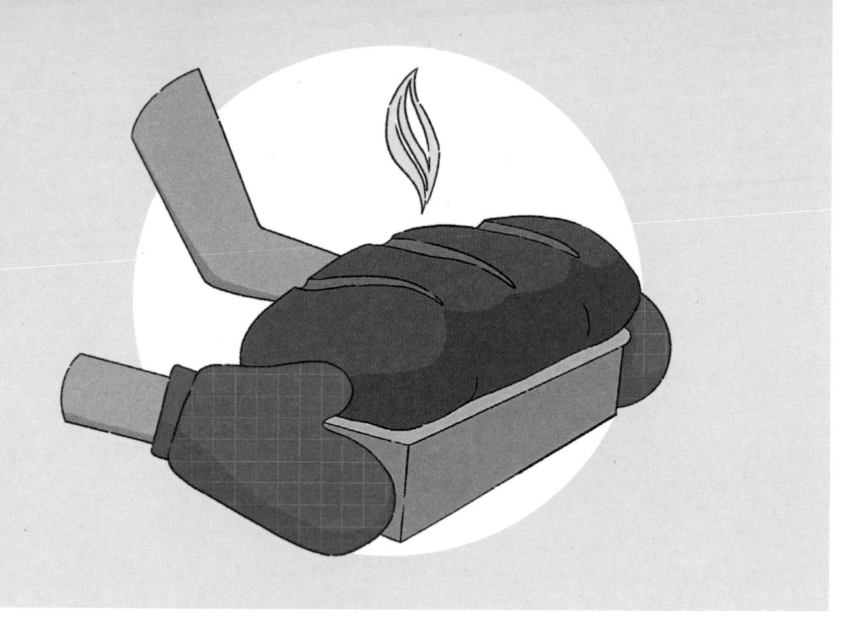

Putting food impacts in perspective

There are strong arguments for eating locally (which usually also means seasonally), but it turns out that transport emissions only account for a small proportion of food's overall carbon footprint.

A comprehensive study done by Dr Hannah Ritchie for Our World in Data, a very helpful project that looks at the biggest problems of the world, and how to make progress against them, found that it's more about what you eat than where it came from. For example, alongside the usual meaty suspects (beef, lamb and mutton particularly), foods such as cheese, chocolate and coffee also have very high associated emissions. Bananas, soya milk and nuts, on the other hand (which we might assume have travelled a greater distance), are comparatively low. For most of the foods considered in the study, more than 80 per cent of emissions arose from land use changes – arising from alterations in soil carbon levels or deforestation, for example – and at farm level: from farm machinery, for instance, or methane emissions from cows.

In short, eating locally only slightly reduces emissions on a macro level – because the carbon footprint of beef or cheese, for example, is principally attributed to how it's produced and not how far it's travelled. The following list shows you the carbon footprint of some typical food items:

Greenhouse gas emissions per kilogram of food product

(kg CO2e – carbon dioxide equivalent)

Beef 60
Cheese 21
Chocolate 19
Coffee 17
Poultry 6
Rice 4
Milk 3
Wheat and rye 1.4
Soya milk 0.9
Root vegetables 0.4

More detail can be found at Our World in Data (ourworldindata.org/food-choice-vs-eating-local)

Things start to get complex when you delve into the nitty gritty of food impact comparisons, and there are caveats. For instance, there are a select group of foods for which transport-related emissions *are* significant: those that are air-freighted. But how do you know which foods come by plane rather than boat? If they have a very short shelf life, like berries or asparagus, and have come a long way, they tend to be shipped by air. While apples don't need to be eaten straight after being picked, raspberries do. So, they need to be speedily shipped to their destination.

Ritchie's study is revealing, and, for cheese and chocolate lovers, so very disappointing! But keep in mind it focused on emissions without touching on the benefits to nature or local communities that keeping it close can provide. Smaller-scale farms that use sustainable agricultural methods are often richer in plant and animal biodiversity, and with almost half of biodiversity already lost in Britain since the industrial revolution, preserving and protecting what remains is vital.

So, what should you do? Eat more plants and plant-based foods; familiarise yourself with high-carbon impact foods; eat locally and seasonally (especially foods that would otherwise be air-freighted: berries, asparagus and green beans, for example); eat organic when you can; and try not to waste food.

And, importantly, don't stress. Supermarkets are beginning to trial traffic-light labelling, which scores food based on environmental impact. If rolled out on a broader scale, this kind of system could make it much easier to choose quickly and easily the foods that are kindest to the Earth, without needing to have a PhD in the subject.

Positivity Pause

Someone told me once that when it comes to intelligence or general capability, you are the average of the five people you surround yourself with the most. It's a fairly rough concept, or perhaps my recollection of it is. But the general principle seems sound: the people I choose to spend the most time with (barring my four-year-old, bless her) represent a kind of wider snapshot of me and my general ability, views and interests.

Presumably, among your five closest friends or family members, there are reciprocal factors at play. Hanging out with your philosopher friend, for example, whose capacity for deep and meaningful discussions about logic and being adds richness to your life. And perhaps your lighter approach to living – maybe you often drag them to parties and festivals – brings a *joie de vivre* that your pensive friend appreciates.

Now, think about the five media sources you frequent the most. It's not a perfect comparison, as newspapers, magazines and online news sites are not people, but stay with me. What are these media sources giving you? Are they reflective of what you want your general outlook on life to be? Do they represent how you want to see the world? Just like you and your philosopher friend, where energies and opinions are likely mutually absorbed, however unconsciously, it stands to reason that what we read is similarly internalised.

As humanity faces what feels like an endless march of crises – around the cost of living, energy, the environment, war, famine, or drought, for example – we need a positive, constructive yet realistic outlook. We also need to move away from the 'bad news bias': the tendency for mainstream media to emphasise what's going wrong and not what's going right.

Five media sources that focus on solutions

- **Reasons to be Cheerful** 'A non-profit editorial project that is a tonic for tumultuous times.' **reasonstobecheerful.world**
- **BBC Future** 'We shine a light on the hidden ways that the world is changing – and provide solutions for how to navigate it.' **bbc.com/future**
- **Positive News** 'Pioneers of constructive journalism – a new approach in the media, which is about rigorous and relevant journalism that is focused on progress, possibility, and solutions.' (Full disclosure: I work for them.) **positive.news**
- **Hothouse Solutions newsletter** 'There is no one solution to climate change. But there are many.' **hothouse.solutions**
- **5 Media** 'We tell powerful stories about sustainable initiatives.' **fivemedia.com**

Leftovers and keeping things fresh

Like lots of parents from her generation, my mum is a leftovers queen. She tucks away the smallest amounts of food – a third of a sausage, a quarter of a sandwich, two segments of apple – and has a very relaxed approach to use-by dates. A few spots of mould on the Cheddar? No problem, cut them off! Lettuce looking brown around the edges? Get your knife and make a salad out of the good bits!

I don't want to advocate for eating food that is past its best. Government guidelines advise not to consume food after its use-by date (use-by dates are about safety, whereas best before dates are about quality; food past its best before date can still be eaten – just do a sniff test first) but there are lots of handy tricks to keep things fresher for longer.

Avocados and citrus fruits

These fruits have a particularly high environmental footprint, so try especially hard not to let them go to waste. Put them, uncut, in a bowl of water in the fridge. You'll be amazed at how long they last.

Spring onions, celery, asparagus, kale, fresh herbs

Stand in a cup of water in the fridge. Herbs can also be wrapped in a damp tea towel. Many cut fruits and veggies, such as carrots, potatoes, apples and beets, will be happy in a bowl of cool water if you need to use them later, too.

Berries

There's nothing better than in-season British strawberries, but, as we've discussed, berries don't hang around for long. A simple trick to prolong their edible life is to dunk them in a bath of one part vinegar, three parts water. This will stop mould from forming so soon. Be sure to dry them thoroughly, though, and place in a dish with plenty of room for them to breathe.

Freezing

Making best use of your freezer is one of the easiest ways to prevent food waste (not to mention save money). Here are a few things you probably didn't realise will freeze well.

Milk
Hard cheese Grating beforehand makes it easier to use handfuls as and when you need.
Eggs Whisked, not in their shells!
Bananas (chopped) You can make a delicious banana 'ice cream' by cutting into chunks and blitzing with a dash of milk and any other tasty extras. I like to add a dollop of peanut butter, hazelnut chocolate spread or any other fruit I have that needs using.
Onions and spring onions (chopped) No need to cook first.
Vegetables, including potatoes (chopped) Lightly blanch them first.
Yogurt
Cooked rice
Wine Pour into ice cube trays and add to recipes when needed.
Fresh and dried herbs
Avocado Peeled and cut into chunks.
Butter

Sustainable sipping

When we talk about climatarian diets, or the carbon footprint connected to what we fuel our bodies with, we tend to focus slightly more on food than drink.

On the juice front, if you're looking for brands that get good grades for their environmental credentials and ethics, Ethical Consumer can recommend some. Human rights and labour abuses are common on fruit farms (particularly Brazil's orange groves) so opt for Fairtrade or organic as a general rule, if you can.

On the dairy side of things, this accounts for over a quarter of the total carbon footprint of typical EU diets. Indeed, we're a latte-loving, cheese-craving, ice cream-infatuated continent. But interest in plant-based milks is on the rise. Surveys have found that one-quarter of adults drink non-dairy milk alongside dairy, and one-third of 16–23-year-olds are choosing plant-based too. But how do the two compare for their environmental impacts?

Dr Hannah Ritchie has done a helpful comparison. Looking at land use, greenhouse gas emissions, freshwater use, and eutrophication (where ecosystems become polluted with excess nutrients), she found that cow's milk has much higher impacts in all of these areas. So, try to go with plant-based, at least some of the time. Some people find it difficult to part with dairy milk in their hot drinks, but you'll barely notice the difference when using plant-based in most other ways. Crunchy granola with oat milk and a handful of fresh fruit is a delicious alternative.

Not sure which non-dairy milk brand to go for? Again, Ethical Consumer is a great resource, with recommendations for brands based on their eco and ethical credentials.

Biodiversity at the bar

Organically produced wines are not too difficult to find. Some supermarkets even have own-brand ranges. And there's a whole world of natural, biodynamic and otherwise environmentally conscious wineries to seek out if you're so inclined. But there's less awareness around climate-friendly beer and spirits – not that they don't exist.

Beer

As far as beer goes, the UK is spoiled for choice. There are more than 2,000 local craft brewers, and some are really pushing the sustainability envelope. Look out for brands that use up surplus ingredients in the brewing process, or who make use of spent grain. Being certified organic or vegan is usually a good sign, and in terms of carbon impact, generally the smaller and more local, the better. A handful of giant multinationals own many of the most well-known beer brands, and historically they've not been particularly forthcoming about their overall impacts. There's good news for those who love their local, too: a locally brewed cask ale at the pub has nearly a third less CO2e than a long-distance bottled beer bought from a shop. So if you needed a reason to head on out, there it is.

Spirits

The good news is that when it comes to environmentally and ethically centred spirits brands, there's a veritable bevy of bevvies on offer. The bad news is that they're not cheap. From biodiversity-focused Calvados to climate-positive gin, eco-centred booze brands are popping up all the time. Look out for distilleries that prioritise sustainable agricultural methods and those that are taking measures to reduce their carbon footprint. Consider giving such climate-friendly indulgences as a very special gift.

Planet-friendly partying: how to put sustainability at the centre of your celebrations

Holiday seasons and parties are a break from everyday life – a time to let loose, splash out and splurge on fun extras. But – and don't let the 'but' leave you crestfallen – they can be pretty wasteful.

Here are a few tips for sustainable shindigs:

- Use good old-fashioned regular plates and cutlery instead of disposable ones. You'll need to accept that there will be more dishes to do (sorry!) but you'll save yourself some guilt at all the rubbish you'll avoid. Alternatively, find compostable ones. Be wary of product claims though: only put items with the Seedling label into your garden waste or food waste bin.
- Ask everyone to take home some leftover food at the end of the day (or night).
- Wrap gifts with reusable wrapping paper. You can get it in beautiful patterns from companies such as Wrag Wrap or HappyWrap. Or, if you're handy with a sewing machine, simply make your own. Buying bundles of vintage scarves to use as gift wrapping is another fun idea.
- Buy gifts from local or sustainability-orientated businesses.

- Give experiential gifts.
- Reuse, reuse, reuse – make gift tags from used wrapping paper or cards, keep hold of fabric ribbons and bows, and use cardboard boxes to post things in at a later date.
- Invest in a good-quality, reusable table cloth and napkins.
- If you're giving party bags to children, include things such as seedballs or edible, homemade treats rather than plastic tat.

Ideas for the festive season

- Make your own decorations, wreath or menorah from foraged or upcycled materials.
- Use candles made from natural materials rather than paraffin-based ones.
- Make sure to unplug lights before you go to bed.
- Think carefully about getting a real versus an artificial Christmas tree. According to the Carbon Trust, a fake tree has more than 10 times the carbon footprint of a real one, provided that the real tree is collected by your local authority after Christmas, to be composted or turned into wood chip. If you do have an artificial tree, make sure to use it at least 10 times, to counter its environmental impact.
- Buy reusable crackers or make your own.
- Just give less – most of us end up with more than we need after the holidays.

Eating out

As a Canadian, I feel no stigma about asking for a doggy bag if I haven't managed to finish my meal. Why this is culturally less accepted in the UK, I cannot say. But eating establishments are becoming more climate-conscious, albeit slowly. According to the UK-based Sustainable Restaurant Association (SRA), there has been an increase in the number of diners ordering plant-based dishes from a reasonable proportion of restaurants within their 10,000-member-strong network. And, after some particularly large establishments changed the portion sizes and overall design of a few of their dishes in order to reduce food waste, they were pleasantly surprised to find that consumers didn't protest.

Not sure how to know whether a restaurant is serious about sustainability? Follow these tips from the SRA:

- Transparency around ingredients sourcing is a tell-tale sign that an eating establishment is climate conscious. Restaurants that care about provenance will champion producers and indicate where ingredients came from, on the menu or on their website.
- Check for dishes that incorporate items that are traditionally wasted, such as cauliflower leaves. Using preserves like homemade pickles, jams and fermented foods (such as kimchi or sauerkraut) are also markers of a chef's commitment to feeding bellies – not bins. No restaurant serious about its impact would refuse a request for a doggy bag, either.

- From a social equality point of view, finding out that the tip you left for your excellent waiter just went into the corporate pot might just leave a bad taste in your mouth. Restaurants that treat their staff fairly generally highlight their tipping policy on the menu.
- Knowing what a climate-friendly meal looks and tastes like isn't always easy. Fortunately, there are programmes such as One Planet Plate, in which participating restaurants indicate the most sustainable dish on their menu as their 'One Planet Plate' choice. Check out oneplanetplate.org to find participating establishments near you.
- Also look out for the SRA's Food Made Good rating system, a comprehensive assessment of a restaurant's sustainability credentials. Participating establishments are given one to three stars.

A few more thoughts on food shopping

A great way to reduce your plastic footprint is to shop at one of the zero waste stores that are appearing on high streets across the country. It takes some forethought as you'll need to come armed with containers to fill, but these shops usually carry all the bulk dry goods you'll need, plus things like sauces, spices and household cleaning products for refill.

With the cost of living and inflation rises of recent years, however, doing most of your shopping at zero waste stores won't be feasible for everyone, as they tend to be more expensive. As a nation, we spend nearly 11 per cent of our income on groceries, and many people can't afford for that figure to increase.

But sustainability is a growing concern for major supermarkets, and the more that consumers pressure them, the faster they will respond. Trials are currently in place for refill stations at supermarkets such as Aldi, Waitrose, M&S and Morrisons, and a review by *Which?* found that the items in these schemes were about 15 per cent cheaper than their packaged varieties.

Laura Parry, who runs Sero, a zero waste store based in Newport, South Wales, says that supporting small shops like hers that focus on sustainability is always going to be better than going with the big guys – but don't beat yourself up if you're not able to do so 100 per cent of the time. 'Because you'll just become disheartened and give up, and that's the last thing we want,' she says. The shop, one of several local businesses based in the grounds of the National Trust's Tredegar House, keeps a running tally of refills. In just over a year, 17,000 containers, jars and bottles were filled up. 'That's how much just choosing to refill one or two things in your household adds up to. Little changes have a big impact,' Parry says.

A few more ideas for sustainable shopping:

- Try to buy loose fruit and veg instead of plastic-shrouded multipacks.
- Plan your meals and do a larger shop instead of more frequent smaller ones throughout the week – you'll end up with fewer impulse purchases.
- Be wary of the 'reduced' section. It's tempting to load up on items that are heavily discounted but they're usually very close to their use-by dates. Only buy if you're sure you'll eat it quickly – or freeze, of course.

Links

recyclenow.com/how-to-recycle/understanding-recycling-symbols

Fashion

The most sustainable washing machine, piece of furniture, kettle, phone (the list goes on) is the one you already own. The same goes for clothes. But nobody wants to keep wearing the same fusty jumper until the end of time. Among my friends and I – and I know we're not alone in this – marrying up how to be stylish while not compromising our values is one of the biggest sustainability quagmires. The key, I think, is progress – not perfection.

So, what does that look like?

None

Admittedly the least fun option, but: just don't buy it. Marie Kondo, the household decluttering guru, says that the things we own should 'spark joy'. Think of that next time you're in a clothes shop, or on the verge of buying a very trendy hairband or slogan t-shirt. Will it still be sparking joy a year from now?

You could also stretch from ***none*** to simply 'less'. The general rule is that you need to wear something at least 30 times to counter the waste and emissions that were created in producing it. That's once a week for seven months. I don't know about you, but seven months does not seem a long time to hold on to a garment, and especially one that is good quality. So, I'd challenge you to wear it at least 60 times. And, speaking of wearing joyful clothing, you could consider getting your 'colours' done, where recommendations are made as to what hues look best with your hair and skin tone; this will likely stop you buying things that you don't end up wearing.

tip:

Orsola de Castro, co-founder of the sustainable fashion campaign Fashion Revolution, recommends turning a garment inside out and pulling at any loose strings before buying. If it starts to unravel, that means the seams have not been carefully constructed and it will likely not last for 30 wears, let alone 60.

Natural

If you can, opt for natural fibres, such as organic linen, organic hemp, organic cotton, recycled cotton and recycled wool, over synthetics. They won't shed microfibres when you wash them, and some, such as linen and silk, won't pill.

tip:

Pilling happens more with knitted than woven fabrics. Although there are no hard and fast rules around which fabrics do or don't pill, a general rule of thumb is to choose tightly woven fabrics and avoid fabric blends. Items made from single component fabrics (100 per cent cotton, for instance) are more breathable than synthetics so you'll wash them less, which means fewer pills.

Neutral

From extending the lifespan of your clothes to buying secondhand and vintage, there are lots of ways to neutralise the effect that fashion has on the planet. The art of slow fashion – a careful, thoughtful and probably more stylish approach to how we dress ourselves – is the antithesis of fast fashion. Slow, sustainable, neutralising – whatever you want to call it, what matters is that you start somewhere. So here are a few guidelines:

Make your clothes last longer

Natural fibres or not, the most important thing is to care for the garments you have so they stand the test of time. Try not to wash your clothes too much – experiment with spot cleaning or wearing slips under skirts so they touch your skin less. You can also get armpit shields – cotton half-moon pads that you pin into the underarms of special garments that you want to preserve.

Amy Winston-Hart, owner of UK-based shop Amy's Vintage, has a few more strategies:

- Make an odour-nixing cocktail: mix three parts vodka and two parts water for very smelly clothes; three parts water and two parts vodka for less offensive items. Pop in a spray bottle and mist over your garment, turned inside out. Reapply the next day if there's a lingering odour.
- Put clothes in the freezer for a refreshing sleepover.
- Hang on the line on a bright and breezy day.

Keep your garments going

The equivalent of 60 lorry loads of textile waste is burned or buried in landfill every minute in the EU – a statistic that is almost too mind-boggling to believe. Although it might seem that recycling textiles is the norm, only a quarter of the 5.8 million tonnes that European consumers discard every year actually go for recycling. And, in fact, only about 1 per cent of recovered textiles are actually turned into new clothes. The rest is reprocessed for use in things like mattress stuffing or car seats. This is because much of the clothing people donate or send for recycling is of too poor quality to be re-made into clothing. Additionally, the technology to recycle many of the fabrics and fabric blends we wear is still nascent.

Scientists and innovators are working on ways to make textiles more recycling-friendly and, indeed, to develop technologies to deal with our discarded duds. In the meantime, do what you can to reduce the amount you donate or put in textile banks. Recycling is important, but it doesn't address the root cause of the global fast fashion waste problem.

So, try to keep your textiles in circulation for as long as possible by swapping with friends, selling on secondhand sites, mending and upcycling, using as rags, getting your grandma to refashion into doll's clothes. You get the idea.

Become a secondhand lover – and spread the word

The internet is a smorgasbord of secondhand clothing delights; you just need to make a point of seeking them out. From the usual suspects such as Depop and eBay, to platforms such as Vinted, Rokit, Etsy and Beyond Retro, you'll find used garments galore if you take the time to look.

Whereas shopping secondhand used to have a stigma attached to it, experts predict the size of the preloved fashion market will be twice that of traditional retail by 2030. And people are much happier to buy secondhand than they were in the past, with one poll finding that almost half of Brits were more likely to purchase pre-owned goods than they were five years ago. Some of these shifting attitudes can be attributed to the cost-of-living crisis of recent years and the Covid-19 pandemic – but every cloud, as they say.

Try to keep the social contagion going and blab about your 'new' preloved dress, shirt or shoes. You may find that friends who wouldn't have considered a charity shop dress a few years ago start to quietly come round to the idea.

Be wary of 'sustainable' fashion

The high street chains are slowly, incrementally, beginning to make changes to their practices – introducing 'conscious' collections that incorporate more recycled and innovative fabrics, putting bins in store for used clothing to be deposited and the like. But the industry is still awash with greenwashing, and some genuinely sustainability-orientated brands are eye-wateringly expensive.

It's good to try and choose the lines or brands that are aiming to do the right thing, but a better port of call is to go with secondhand or vintage. Doing so will mean you're making a small contribution towards lessening the water, waste and carbon impacts of the fashion industry, not to mention its plastic footprint. Fashion accounts for a fifth of the 300 million tonnes of plastic produced worldwide each year.

If it's something extra special you're looking for, like a dress for a wedding, vintage is often the best choice. You can pretty much guarantee no one will have the same outfit, and you'll be wearing a piece of fashion history. A growing number of companies now offer formal wear for hire too. Check out HURR or By Rotation for women, and Peter Posh or the Vintage Suit Hire Company for men.

Positivity Pause

The writer and researcher Britt Wray says that the climate crisis is a mental health crisis. It's an acute statement that, interpreted in a certain way, could make things feel worse: that the climate crisis is almost anthropogenic. Not satisfied to remain materially external, this nightmarish beast now seems to have the power to invade our private selves too. Wray argues, however, that facing up to what is happening, however painful, can be a significant motivator for systems change. And there are signs that people across the country are seeking, and building, therapeutic outlets to discuss the relatively new phenomenon that is climate anxiety.

'Climate cafés' are informal gatherings where people share their thoughts and connect with others over tea and cake. Instead of feeling like they're bringing down the mood by talking about these issues, attendees say having a space to connect with others on shared concerns is a deep relief. They're popping up in ever-greater numbers – at community centres, online and at university campuses.

The University of East Anglia, as part of its sUStain project with the Norfolk and Waveney branch of the charity Mind, offers monthly climate cafés and wider workshops on the concept of 'active hope'. The idea, pioneered by resilience expert Chris Johnstone and eco-philosopher Joanna Macy in their book *Active Hope: How to Face the Mess We're in with Unexpected Resilience and Creative Power*, is about being involved in bringing about what we hope for – not just passively waiting for external forces to do the job.

This idea, that in order to combat feelings of despair, we must take more of a participatory approach to solutions, is also being borne out by a new trend taking hold of high streets. As retail giants abandon shops, community groups are moving in: an old New Look store in Guildford, Surrey, has become a 'climate emergency hub' focused on locally orientated solutions. The centre also aims to help build up mental health and well-being support frameworks. Elsewhere, an empty unit in Stretford Mall, Greater Manchester, has become a sustainable fashion hub called Stitched Up. There people can buy secondhand clothing, learn how to sew and, importantly, connect with others.

Watch out for community-centred, action-orientated enterprises like these where you live. Being involved in solutions might go a long way towards curtailing feelings of powerlessness. As the academic and author Dr Sarah Jaquette Ray says: 'A sense of the collective is probably the most important thing that will alleviate climate anxiety, but also mitigate climate change.'

Technology

Rather than beginning this chapter by painting a harrowing picture of the growing scourge that is electronic and electrical waste – visions of people in developing nations crouched over soups of smoking metal and so on – let's immediately focus on the factual realities of the problem, and what's being done to tackle it.

The facts

According to the UN, e-waste is the fastest growing domestic and commercial waste stream, with recovery, repair and recycling programmes not being developed fast enough to keep up. And the UK is far from a shining example. According to the Global E-waste Monitor, we generated the second highest amount of e-waste per person in the world in 2019: nearly 24kg each. The only country to surpass that was Norway.

Individual spending on tech continues to grow, with research showing that 40 per cent of new purchases are driven by the desire to upgrade. Surveys have also shown that people in the UK are not very savvy when it comes to responsibly disposing of their tech. YouGov research from 2022 found that one in five people were unaware of how to deal with unwanted electronics and devices. And almost a quarter found the recycling of such items inconvenient or confusing. What's more, only 2 per cent of those surveyed said they fixed or took electronics to be mended.

There's also the issue of e-waste being illegally exported abroad. Under EU law, this type of hazardous waste, which

contains substances such as mercury, lead and flame retardants, cannot be sent to non-OECD or non-EU countries. But it still manages to find its way there. A two-year investigation, done by the environmental watchdog the Basel Action Network, put GPS trackers on items such as LCD monitors and computers and sent them off for recycling through local council schemes. But the items were traced to suspected illegal shipments heading to Nigeria, Tanzania and Pakistan. And the UK was the worst offender among the 10 European countries in the study.

The global e-waste problem is multifaceted, with action needed from all sides: government, tech companies and consumers. But small steps of progress are being made.

From the beginning of 2021, tech retailers were obligated to take back used electrical and electronic equipment (WEEE) on a 'one-for-one' basis, with online-only retailers required to do so from 2022. This means that you can bring in your old TV or kettle, for example, when you buy a new one. And retailers with a floor space of at least 400 sq. metres must accept 'very small' items irrespective of purchase. Retailers must also offer information about their take-back scheme at the point of sale, so if nobody mentions it and you don't see a leaflet, do ask. With this expanded network, there should now be more than 10,000 retailer collection points, up from about 400–500 previously. This may make it easier for those who don't have vehicles to return their devices, as retail parks are often nearer to public transport links than Household Waste and Recycling Centres.

Some retailers have started public awareness campaigns to boost understanding of the impact of consumers' purchases and encourage people to recycle and repair their gadgets. And some tech giants are caving in to pressure to make it easier for people to repair their own devices, beginning to answer calls from the 'Right to Repair' movement to make spare parts available to consumers, for example.

On the policy front, EU legislation around chargers for electronic devices, such as phones, tablets, e-readers and headphones, is also imminent. A policy mandating that all chargers are of the USB-C variety is intended to cut e-waste, but some sceptics are worried that universalising how we power up our devices might make it even easier to justify buying more of them. And chargers comprise but a drop in the e-waste ocean. But it is a step in the right direction.

In the UK, the Department for Environment, Food and Rural Affairs (Defra) is in the process of consulting on current

WEEE and battery recycling policies, with a view to proposing updated legislation in 2024 or 2025. The word on the street is that they'll focus on dealing with planned obsolescence, where manufacturers purposely – and irresponsibly – design products with limited lifespans, instead of incorporating longevity through options for repair, for example.

Meanwhile, researchers are hard at work on a citizen science project called the Big Repair Project, conducted through University College London. By surveying consumers about their attitudes towards appliance and electronics repair, the study aims to identify the pinch points, as well as repair 'hotspots': areas where local fix-it economies are thriving.

And speaking of repair and refurbishment – arguably where the main focus should be if we are to build a more circular system – a burgeoning movement of non-profits, businesses and community groups are dedicated to the cause. The Restart Project, a social enterprise that teaches people how to repair their broken electronics and rethink how they use them in the first place, is a great example. Bring any tech in need of TLC to one of their UK-wide 'restart parties' and volunteers will get to tinkering, and show you how too. Or pop along to one of their 'fixing factories' in Camden or Brent in London; these are community facilities that aim to redefine the very meaning of production, where what comes off the 'conveyer belt' are repaired goods and a sense of empowerment.

Society may only be taking baby steps so far when it comes to reversing the tide of e-waste, but the desire for change is there. A survey by the Royal Society of Chemistry revealed that over 60 per cent of people would switch away from their preferred tech brands if rivals were prioritising sustainability. But while we wait for big tech to do the right thing, let's take things into our own hands.

Think twice before you upgrade your phone

It seems that in life, there are people who are into tech, and there are people who are not. Advising the latter group only to upgrade their gadgets unless it's truly necessary is not a particularly big ask. Convincing the former, however, is trickier.

One statistic could go some way in putting things into perspective for those infatuated with upgrade culture: buying a new phone takes as much energy as it does to recharge and operate a smartphone for a decade. This is because most of

the climate footprint of a phone comes from its production: a study from McMaster University in Canada found that mining the rare minerals contained in a phone accounts for 85–95 per cent of the device's total CO2 emissions for two years.

It's sort of like the statement about swapping meat-based dishes for plant-based ones (and how doing so saves the equivalent amount of energy required to charge your phone for a year, if you remember), in that it takes a second to appreciate what it means. You can be sure that being the person who peppers a conversation with foreboding facts won't be effective, but you can lead by example. Or at least be the recipient of your upgrading friend's hand-me-downs. (Or if you're the upgrader, make sure you recycle or pass along your old handset). But if recent statistics are anything to go by, people will, on the whole, start to upgrade less.

Phone companies would have us believe that we need to upgrade our phones every two years, but a UK survey found that people are hanging on to their current models for longer than they would previously. Support from the big phone manufacturers for older models is slowly expanding, but people have also generally been cutting their costs due to the pandemic and economic crisis. Silver linings again.

Fix it

When it comes to phones, people are generally aware that smashed screens can be replaced, depending on how severe the damage is. But did you know that batteries can typically be replaced too? *Which?* Has a useful guide that compares battery and screen replacement costs, from official retailers and third parties. Find the link at the end of this chapter.

For other electronics, it's worth investigating what repair options there are in your local area, be they restart parties or local businesses. You can also consult ifixit.com for guides on repairing everything from medical devices to game consoles, cameras, microwaves and household tools.

Consider also joining the Right to Repair movement, an international effort to secure repair options and prevent limitations around it. Across Europe, the winds of change are blowing with regard to legislation. In France, for example, repairability score labels are now required on five categories of consumer goods, including electronics. Encouragingly, most French consumers in a survey said they were using the labelling to buy more repair-friendly goods. In the UK, progress has been made around white goods but as far as electronics go, it's up to companies to honour this right.

Catalyse further action by joining the EU's Right to Repair movement at repair.eu.

Buy refurbished or secondhand

It's a bit of a no-brainer but worth a reminder that there are plenty of quality refurbished electronics and electrical goods out there; just make sure to buy them from a trusted source. Unlike secondhand, refurbished items will come direct from the manufacturer or a certified provider and be professionally checked. *Which?* has resources to help, with specific guides for buying secondhand and refurbished laptops and mobile phones.

In terms of white goods or small appliances, you can often find secondhand models on local marketplaces or even in charity shops. Buying from retail giants is not usually a particularly ethical or sustainable choice, but, if you're struggling to find quality secondhand options where you live, they're worth investigating for their refurbished offerings. A quick comparison between one manufacturer's website and its online outlet revealed that the same model of machine was being sold for £250 refurbished versus £379 new. So you'll be saving not just carbon, but cash too.

Recycle, donate or sell, but try not to hoard

We hold on to our old tech for a variety of reasons: because we think it might come in useful, there may be concerns over the data that's on it, it's too much of a bother to deal with old stuff, or we just don't know what to do with it. Whatever the reason, the UK is hoarding around £7 billion worth of used electronics, which experts say is hindering the development of a circular economy for e-waste. Reports have also found that about 300,000 tonnes of electricals are being thrown in the bin every year by UK households, with huge amounts of valuable metals being wasted.

But value is not just being lost through irresponsible disposal: inefficient recovery methods are also partly to blame. According to Material Focus, the non-profit behind the Recycle Your Electricals campaign, less than

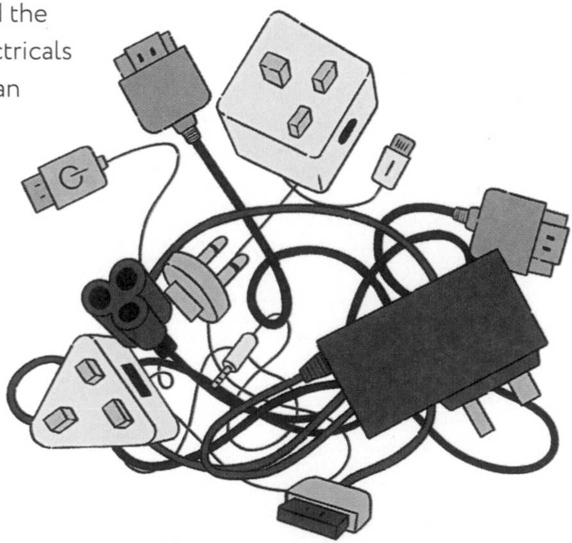

1 per cent of all rare-earth elements that are in electricals are being recycled. Precious materials such as gold, platinum and silver are slipping away, or being kept in cupboards, when they could be put to good use.

In fact, if all the materials from waste electricals were recovered and recycled, we could make 11,785 catalytic converters from the platinum, 500,000 wedding rings from the gold and 2,661 wind turbines from the neodymium.

Improving recycling technologies will lie in the hands of industry, but consumers have a huge role to play in helping to make best use of old electronics and electricals.

Uswitch has a helpful guide on how and where to recycle your old phones, as well as a database that allows you to check whether items, from curling tongs to old Mp3 players, are recyclable. Use the Recycle Your Electricals campaign to find your nearest drop-off point and check with local council kerbside collections too.

Links

recycleyourelectricals.org.uk

which.co.uk/reviews/mobile-phones/article/mobile-phone-repair-cheapest-way-to-fix-iphone-or-android

therestartproject.org/parties

uswitch.com/mobiles/guides/mobile-phone-recycling-explained

Positivity Pause

In behavioural science, there is something called the 'intention-action gap'. In simple terms, it refers to what people say they plan to do, and what they actually do. It's a significant hurdle, particularly when it comes to sustainable habits. Research has shown that 92 per cent of people want to live a sustainable life, but only 16 per cent are making changes to realise that goal.

So how can we close the gap?

The book *Atomic Habits*, by James Clear, has some answers. In order to adopt a new habit and make it stick, we need to follow four principles.

First, make it obvious. Clear says, 'environment is the invisible hand that shapes human behaviour'. What's around us, he argues, is often the most important thing in terms of how we act. For instance, products put at eye level on supermarket shelves are purchased more than those in harder-to-reach spots simply because they're in an obvious place. Do you keep forgetting to take your reusable bags with you to the supermarket? Put them right by the door next to your keys and keep some in the boot of your car or in your handbag.

Next, make habits attractive by getting your brain to

associate them with a reward. Perhaps if you've managed not to forget your cloth bags for three months, you can reward yourself with something you love, like getting a pedicure or a massage. You could also join a culture that sees your old habit as unattractive. For example, zero waste stores promote reusable receptacles, so asking for a disposable bag in one of these establishments would be undesirable.

Third, make it easy by removing any friction. Or add friction to the undesirable habit. In the reusable bag example, some friction is already built in: you have to pay 10p for a disposable one. And in terms of opting for no bags at all when getting food delivered, some supermarkets make it the default to go 'bagless', thereby removing the added step of you having to tick a box to do so.

Finally, make it satisfying. Clear says that the first three principles make it more likely that a behaviour will be performed. But making it satisfying will increase the odds that the behaviour will be repeated again and again. Introducing 'habit tracking' can help to up the satisfaction element. For example, for every 10p you save using a reusable bag, you could put, say, £1 in a piggy bank. If you use two reusable bags per week for a year you'll have over £200 in reward money, plus you'll enjoy the ritual activity of plonking a coin in your jar.

Another bit of advice to ensure that habits stick is simply to do them over and over and over again. 'The more you repeat an activity, the more the structure of your brain changes to become efficient at that activity,' says Clear. Over time, your sustainable habit will become something you won't need to consciously think about anymore; it will simply be part of your routine.

Transport and Travel

The UN has labelled the 2020s the 'decade of action', and when it comes to the greening of transport, the pronouncement certainly rings true. The remainder of the 2020s will see the near extinction of internal combustion engine (ICE)-powered vehicles as bans on their sale draw nearer, people seek more environmentally friendly transport and electric vehicle (EV) costs come down.

At some point in the next few years – quite excitingly – we'll reach a tipping point: the adoption of EVs will speed up and they will begin to outnumber ICE-powered cars on UK roads. Some even argue that we've reached that point already. A 2022 survey, by professional services firm EY, found that 49 per cent of UK consumers want an EV, rising from 21 per cent who said the same two years before. It points to increasing confidence that electric cars are the better choice, as anxieties around range are slowly allayed through improved battery technology (the average range of an EV is around 200 miles but in the next few years that will rise to 500 miles), and government restrictions come into force around heavily emitting cars. Access to charging infrastructure remains a sticking point but there is lots going on behind the scenes to boost that up.

It's a once-in-a-century transition, this wholesale shift away from conventionally powered cars, and worthy of massive celebration.

The transport pyramid

A shift is also taking place around what it means to be efficient in our mobility habits. With more people working from home following the pandemic, the top of a newly imagined travel hierarchy is actually not a form of transport at all: it's digital communication. This virtual travel is a neat ***none*** within our three ***Ns*** guiding triumvirate.

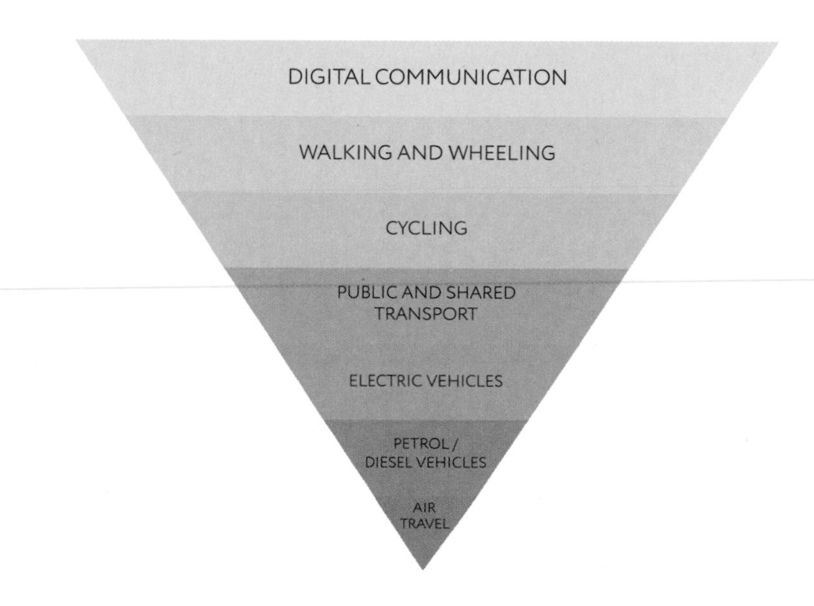

For more details see: energysavingtrust.org.uk/an-introduction-to-the-sustainable-travel-hierarchy

You could interpret the next two levels in the hierarchy, walking and wheeling (the term used for wheelchair users), and cycling, as natural forms of transport too, since they're powered simply by the food we eat. Which, as it happens, has an impact on how eco-friendly these modes are. Mike Berners-Lee, author of the book *How Bad Are Bananas? The Carbon Footprint of Everything*, has calculated that cycling a mile powered by bananas equates to 40g CO2e, while cycling the same distance fuelled by cheeseburgers means 310g CO2e. Food for thought.

With the advent of increased remote working, studies show that people are exercising less and eating more. But incorporating just a few small walks a day can make a huge difference to overall health and contribute to taking cars off the roads. Doing just 10 minutes at the beginning of the day, at lunch, and in the evening will add up to 210 minutes a week, well above the 150 minutes of 'moderate to vigorous physical activity' recommended by the UK's chief medical officer. For optimum health, try to make at least one of those walks brisk.

In terms of cycling, the benefits are quite clear-cut. As an emissions-less form of transport (when you don't count the bananas or cheeseburgers, or its embodied carbon), it's hard to go wrong. However, the sector as a whole lags behind when it comes to environmental, corporate and social responsibility. So do try to support brands that have so far made a modest effort in acknowledging and mitigating their impacts. Of course, buying secondhand, and maintaining the bike you already have, should take precedence over buying new.

Options with neutralising features come up next on the pyramid: public and 'shared transport', which refers to car clubs. According to the Energy Saving Trust, the average club car produces 26.5 per cent fewer emissions than the average private car. Plus, each of these shared vehicles takes around 18.5 cars off the road. You'll need to do some research to see if a car club membership works for you – it helps if the pick-up and drop-off points are in handy locations – but it's well worth investigating. I recently became a member of a car club in Canada, which I use when I'm on extended visits there, and it is wonderfully convenient. I can borrow a car for just a few hours if I need to, most of the models are hybrids, and when a petrol top-up is needed, it's included in the price. Search for your nearest car club using sites such as como.org.uk.

Car sharing falls next on the hierarchy, alongside EVs. A slightly different concept from car clubs, where fleets of vehicles are owned and managed by companies, car sharing refers to individuals hiring out their private vehicles by the hour or day, rather like Airbnb but for transport. Platforms Turo, Hiyacar or Getaround are some examples.

A note on train travel

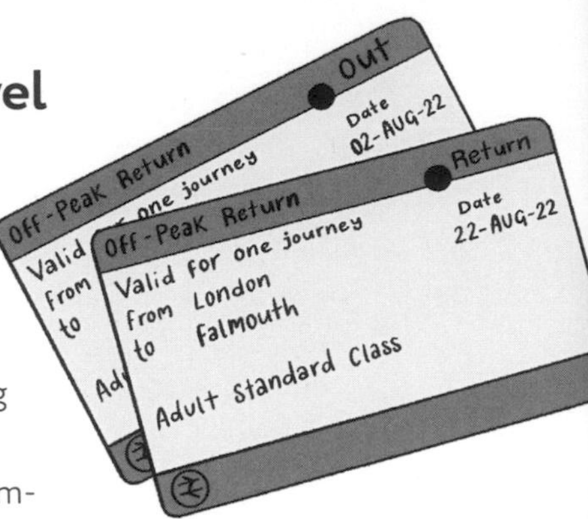

Whether you're commuting or taking longer journeys, train travel is up there as one of the least polluting forms of transport after walking and cycling. In short, taking a train instead of a car for medium-length distances would cut your emissions by about 80 per cent. And opting for a train instead of a domestic flight is even better, slashing your emissions by about 84 per cent. These figures can vary depending on how the electricity grid is powered – for example, in France the grid is 90 per cent fuelled by low-carbon sources, so emissions savings are even more significant there.

tip:

Do you want to train it but find the prices steep? Check out splityourticket.co.uk to search for cheaper fare options. The database searches for ways to get to your chosen destination that involve a 'split' – a pit stop part-way through the journey where you usually don't have to get off the train.

Is it worth buying an EV now?

If you're not sure whether it's a good time to invest in an EV, the answer is not black and white. It may be that holding on to your current vehicle, provided it's not a gas guzzler, is the best choice for now: remember that there's a huge environmental impact embedded within many large, manufactured items. To make an EV, the average estimated carbon footprint is 10–11 tonnes of CO2e, while for a petrol or diesel car it is 6–7 tonnes, according to Ethical Consumer. This is because it takes quite a lot of electricity to make the battery. As we increasingly shift to a renewables-powered grid, here and abroad, this estimation will come down. And when you factor in the emissions saved through clean driving, you'll make back the extra manufacturing impact in about two years. So, it's definitely worth investing when the time is right for you.

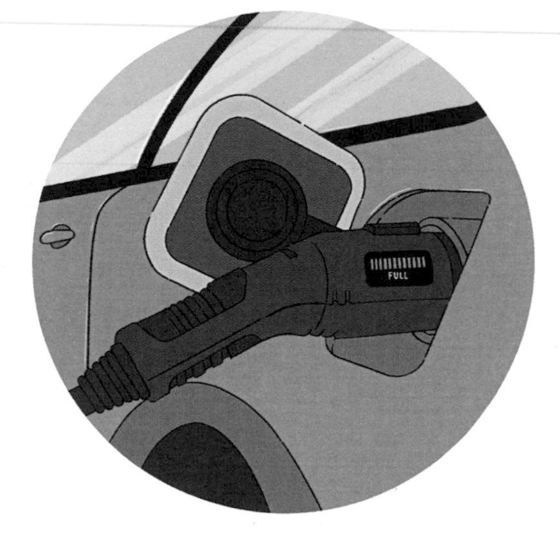

Depending on where you live, you might not have access to charging infrastructure yet, although do be aware that if you live in rented accommodation or a flat, there is some government help available. You can claim back 75 per cent up to a maximum of £350 on a smart home charger.

The network of public charging is also steadily expanding, with certain retailers promising to install hundreds of chargers at stores across the country in the next few years. Although there's still a long way to go, we're headed in the right

tip:

If you're weighing up whether to get a hybrid or pure electric car, Ethical Consumer recommends the latter choice. A wide-ranging study that looked at real-world driving of around 100,000 plug-in hybrids found that emissions were two to four times higher than those claimed by official estimates. This was because people weren't charging them very often and were therefore relying too much on the internal combustion engine.

direction: from 2019 to 2022, the number of public chargers across the UK rose by 129 per cent, and London has the most public rapid charge points by volume of any European city.

It would be remiss to discuss electric vehicles and not acknowledge the wider impact inherent in mining the metals needed to produce them, however. Lithium, nickel, cobalt and manganese are required to manufacture EVs' lithium-ion batteries – resources that, given the surge of demand for EVs that's right around the corner, are very actively being exploited and prospected for. Mining involves infringements upon the rights of indigenous groups in many cases, as well as the destruction of pristine natural landscapes.

These implications are no small issue but sticking with emissions-belching vehicles isn't an option. So, what should you do? Choose automakers that have the best track record in terms of their environmental, social and political policies. Turning again to Ethical Consumer (a yearly subscription to which is not too costly and opens up a huge range of helpful guides and research), look for brands that top their EV shopping guide. At present, however, none of the available options have been given their 'best buy' stamp, which they reserve for companies with top marks across all categories.

Travel

When I was 15, I took a Greyhound bus from Ottawa, Canada to Port Charlotte, Florida, with my sister, who was 18. It took three days and two nights, and we made about 20 transfers. I remember letting my ticket unfold from my raised hand, each segment of the journey accordion-ing out until the whole thing nearly reached the floor.

My sister and I had saved up money from our after-school jobs to take this trip to visit our grandpa and step-grandma. It was not a very efficient use of travel time for a week-long March break, and what lay at the end of our journey wasn't particularly rock'n'roll. My grandparents lived at Maple Leaf Estates during the winter months: a well-kept, gated retirement community with extracurriculars such as pottery lessons, shuffleboard (an outdoor game somewhere between curling, lawn bowling and croquet) and reclining by the shared swimming pool. Not exactly what spring break dreams are made of, but we didn't have much money and we wanted to go on an adventure.

I'm sure that if my sister and I had had enough cash, we would have taken a plane, and our journey time would have been cut dramatically. At that point in our young lives, and because it was a different time (the early noughties), environmental considerations didn't play a part in the decision whatsoever. Then, we were part of the vast majority of the world's population that didn't ride on planes. Indeed, only 2–4 per cent of people fly internationally (as per 2018 figures), principally because it's not economically feasible.

I have more money now and I live a few thousand kilometres away from most of my immediate family, so I fly to Canada every now and then to see them. I love taking these trips back home, but I don't feel great about them because I know that the climate impacts of flying eclipse other transport emissions.

For context, a return flight from London to Singapore stacks up at around 3 tonnes of CO2. According to *Which?*, that's about how much it would take to heat a family home for a year. And the total average footprint of a person in the UK is 10 tonnes per year (which is about double the international average, by the way).

I feel guilty for flying, because I know it has a huge climate impact, and I know I'm not alone in this quandary. How can we square such decisions with ourselves?

Here's my hierarchy of transport, specifically applied to flying.

First level: don't fly

If giving up flying is something you feel you can do, I applaud you. Or if you can't afford to fly, consider this a blessing in disguise. You'll probably have more fun taking the long way anyway, like my early-noughties teenage self. I saw the inside of every major coach station, from New York City to Baltimore, Maryland, and Fayetteville, North Carolina to Jacksonville, Florida – and I didn't eat a vegetable for three days (Greyhound pit stops are devoid of such things). But I had the best time.

Second level: fly less

If you used to take one family holiday abroad a year, perhaps that could be reduced to one every two. You could take the Flight Free UK pledge, where people commit not to fly for one year – it's impactful but it doesn't box you into 'forever'.

Brits fly abroad more than any other nation. And yet there are fabulous opportunities for staycations here, from glamping at working farms to mini-breaks at historic sixteenth-century farmhouses. The National Trust has a selection of cottages available for booking and the benefit of staying with them is that all proceeds go back into their conservation work.

'Slow' and flight-free travel services are also popping up. There's Byway, a holiday planner that lets you build a flight-free trip, personalised around your interests, whether that's a coastal jaunt to Cornwall or a trek to the South Downs. Wilderness Scotland, meanwhile, specialises in outdoor escapes and can let you know the carbon footprint of your trip. On average, one of their wilderness trips equates to 21kg CO2e per traveller per day, while a Caribbean cruise stacks up at 445kg CO2e per traveller per day. If you do decide to do a European jaunt, there are lots of flight-free options such as the Eurostar, the shuttle or a ferry.

tip:

Find vegetarian or vegan-friendly accommodation and eating establishments at happycow.net.

Third level: fly, but offset it

Offsetting has developed a bit of a bad rap because it doesn't solve the root of the problem and can be seen as a cop-out. But this book is not about trying to be a sustainability perfectionist. It's about recognising the impact you have and doing your best to make that impact as positive as possible.

The general principle of offsetting is that you calculate the carbon impact of a given journey and then fund a project that reduces emissions by the same amount.

Some airlines give you the option to offset through their own schemes at the point of purchase, but, according to *Which?*, their calculations aren't always very reliable. It recommends that the most accurate way to ensure that emissions expended are comparable to green actions funded is to calculate them yourself using the International Civil Aviation Organization's (ICAO)'s calculator. Then, use the non-profit Atmosfair's 'offset desired CO2 value' option to input those emissions and pay for offsetting.

The money that Atmosfair collects goes to developing renewable energy technologies in places where such things are basically non-existent – primarily developing nations. This way, it says, CO2 emissions are saved through the fact that fossil fuels are avoided in these areas.

In a perfect world, green-powered planes would whisk us to far-flung destinations, or at least the technology for such renewably charged flying machines would be just on the horizon. But at present, these technologies are still nascent. Professor Pericles Pilidis from the Centre for Propulsion Engineering at Cranfield University says hydrogen planes, for example, are about 10 years away from being viable. Offsetting represents a fairly long-term stopgap.

As Atmosfair says, 'Offsetting cannot solve the problem of climate change since it does nothing to change the actual source of CO2. It is a necessary second-best solution as long as the best solution does not yet exist.'

So, try to stick to the flying hierarchy and take inspiration from others who have ditched air travel. Helen Coffey, the travel editor of *The Independent*, gave up flying for three years and captured her experience in a book: *Zero Altitude: How I Learned to Fly Less and Travel More*. She describes overnight ferries and sleeper trains as positively exotic, and talks about the renewed sense of adventure that slower and more thoughtful travel can imbue.

I, for one, will always remember winding down a curving highway slip road somewhere in New Jersey, a golden sunset glow reflecting across the Hudson River on my left, and 'Ms Jackson' by Outkast tinkling in my ears. The Manhattan skyline was glinting on the other side of the river, and it was one of those slightly out-of-body moments – a memory I'll never forget.

Links

como.org.uk/shared-mobility/shared-cars/where

nationaltrust.org.uk/holidays

byway.travel

wildernessscotland.com

flightfree.co.uk

which.co.uk/reviews/airlines/article/airlines/carbon-offsetting-how-to-reduce-the-impact-of-flying

icao.int/environmental-protection/carbonoffset/pages/default.aspx

atmosfair.de/en

Nature

On 23 March 2020, the UK shut down. Following then-Prime Minister Boris Johnson's announcement a few days prior, most businesses closed their doors, people stopped going to work and school, and life came to a grinding halt. It was a dark time, but the next few months would reveal something profound: humans need nature.

This was true in deep, intrinsic, unapologetic and practical ways. And we had perhaps been taking that for granted.

With public parks and children's playgrounds closed, and only one solitary exercise excursion per day permitted, access to nature became a privilege and not a right.

And so, the nation started gardening. Granted, it was something to keep us busy when socialising was off the cards, but a huge number of people also found that it helped to relieve stress and improve well-being and that it gave them a sense that they were taking control, in a small way, over food security. Even those who didn't have access to a garden, but instead tended to indoor plants or window boxes, found solace in the hobby. A survey conducted by the charity Mind found that since the pandemic, gardening has boosted the mental health of over 7 million people across the country, and 43 per cent of those surveyed said that indoor planting had the same effect. And with sales of seeds continuing to burgeon post-pandemic, this group of newly minted green thumbs may keep their hands in the soil for the longer term.

But the surging appreciation for nature during the pandemic wasn't limited to gardening. Following the reopening of parks and beauty spots, as well as the relaxation of rules around socialising outdoors, data from the Office for

National Statistics (ONS) found that the number of people flocking to these spots was much higher than previous years. Cornwall saw a 280 per cent increase in visits to parks between January and September in 2020, for instance.

Liz Ware is the founder of the charity Silent Space, a project where participating gardens reserve a particular spot on their grounds for visitors to be silent. The idea is to allow people to take a few moments to switch off from technology and the outside world, and simply absorb nature. With over 70 sites mostly in the UK, Ware says that some couldn't run their space over lockdown, but that since reopening, 'we no longer have to explain why we're doing what we're doing. Everyone gets it!'

Indeed, Covid-19 put a spotlight on the value of nature – but, pandemic or none, business, government and society as a whole are placing greater emphasis on the role that it has in improving well-being, not to mention in maintaining a liveable, sustainable world.

From the rise of social prescribing, where healthcare workers recommend things such as gardening or cookery instead of conventional therapies, to the surge in rewilding, where areas of land are left to return to their natural state, recognition is growing around the symbiosis between humans and nature. Subject to regulator approval, it will even be part of the national curriculum come September 2025: students will have the option to do a natural history GCSE, giving them the chance to improve nature literacy and learn about British wildlife.

Meanwhile, the UK government has committed to

protecting 30 per cent of land and marine areas by 2030, through the High Ambition Coalition for Nature and People. It's one of over 100 countries to make the pledge. And targets are in place to make available £500m of investment per year by 2027 to support nature recovery, increasing to at least £1bn a year by 2030. Some projects in the stream are already underway, such as the building up of natural flood risk management systems by the National Trust in West Yorkshire, and wetland creation in Norfolk.

It's all encouraging, and very much needed. The UK is one of the most nature-depleted countries in Europe, with more than 40 per cent of native flora and fauna species having considerably decreased since 1970, according to the National Biodiversity Network.

But what can you do on an individual level? Lots, in fact.

Sustainable gardening

You may think that gardening is inherently a sustainable thing to do – and I would agree with that. But you can make your garden even greener by following these tips.

Go peat-free

Peatlands are wetland ecosystems that occupy a significant 12 per cent of land area in the UK. Their boggy conditions produce peat: partially decayed plant material that is special for a few reasons.

Through photosynthesis, plants in peatland capture CO2 from the atmosphere. They then lock it in because plants that grow in these waterlogged conditions never fully decompose. This would be quite a helpful phenomenon given how emissions-happy society currently is. Unfortunately, however, peatlands worldwide are a net contributor of GHGs because of how they have been (mis)managed. Once peat is extracted or peatlands are drained, the carbon is released.

Globally, peatlands hold twice as much carbon as the world's forests. They also support flood management and are a habitat for many rare and diverse species. They're so unique and biologically important that they're sometimes referred to as the UK's rainforests. It's therefore especially important to keep them intact.

Buy only peat-free compost and plants that have been grown peat-free. You may need to enquire at your garden centre about this but don't be shy – letting them know people want this will help to speed up action. The Wildlife Trusts' website has a list of major retailers that stock peat-free.

Sow seeds of hardy plants directly into the soil so you don't need pots or compost.

Experiment with alternatives such as bark chippings, wood fibre or coir (a fibrous material extracted from coconut husks).

You can also, of course:

Set up a home compost

Composting is one of those things that sparks rather strong opinions. 'It will attract rats', they say! 'It will smell', they protest! When I was at university, my flatmates and I were disappointed that there was no food waste collection where we lived – so we decided to start composting on our balcony. We got a big plastic bin, some 'red wiggler' worms and began chucking our food waste in it. Every so often we'd stir it with a big wooden spoon. It was a rudimentary system, but it seemed to work (except for avocado skins; worms have a hard time with those). It was also a great conversation starter at parties.

I recommend doing more research into it than my flatmates and I – you can get purpose-built systems or make your own. But don't write it off before doing your due diligence. A properly managed heap shouldn't attract unwanted pests or odours. Nature presents all sorts of perfect solutions to humanity's worldly problems (i.e., unwanted food peelings) – we need to embrace them more!

According to Recycle Now, a campaign across England and Northern Ireland, composting at home for one year can save the equivalent amount of GHGs produced by your kettle in a year, or your washing machine in three months. It will help you prevent sending food and garden waste to landfill, especially if you don't have a local collection, and it will enable you to produce rich, nutritious soil for your garden.

If you have enough space, you can create a two-bed system: one main bed and then another to turn the heap into. You can also buy a bespoke bin that doesn't require turning. Organisations such as the National Trust, the Eden Project and Recycle Now are reliable sources for information on how to set up your compost heap. And if you don't have room for a traditional system, vermicomposting is a much more space-efficient way to turn your food scraps into black gold. The Royal Horticultural Society (RHS) has an excellent beginner's guide to worm composting.

Harvest rainwater

Installing a water butt to collect rainwater from your gutters is a fairly simple one-off change you can make and has lots of benefits. Rainwater often has a lower pH than mains water, which plants like. According to the RHS, the minerals found in mains water, especially in hard water areas, can raise the pH of the root zone, affecting nutrient availability. It takes energy to treat mains water, so by tapping into it less, you'll be lowering associated emissions. Water barrels also help to prevent flooding as the load on drainage systems is reduced.

If you don't have room for a water barrel, you can conserve water by using 'self-watering' planters or putting trays beneath your pots. Adding mulch around plants also helps the soil retain moisture, and simply using a watering can instead of a hosepipe reduces water use.

Pledge to go mains water-free through the RHS's 'Mains to Rains' campaign.

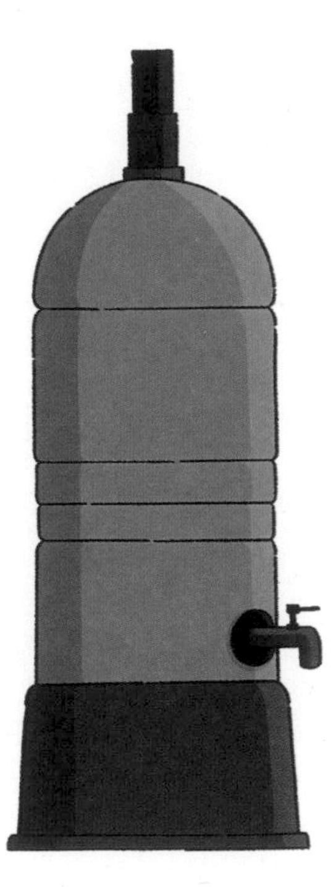

Plant for pollinators

From mowing less to planting specific species, there are lots of ways to create a garden that biodiversity-boosting bees, butterflies and insects will love. Artist Daisy Ginsberg has created an algorithm that helps you design a pollinator-optimised garden. Visit pollinator.art to input the specifics of your space and receive customised planting instructions. Or follow the Wildlife Trusts' simple instructions for creating a 'nectar café', which include ensuring you have a selection of flowering plants for all seasons, choosing plants that have a simple (often flat) structure and planting herbs, which insects love.

Go wild

Over the last decade, rewilding has gained momentum, the idea being that nature is best left to itself. Humans can lend a helping hand by reintroducing species that have dwindled, removing dykes and dams, and curtailing active management of species. We should then lay off and let nature self-govern and self-regulate back to health. Rewilding Britain, which is on a mission to restore 30 per cent of Britain's land and seas within the decade, says that rewilding works best on a large scale, on upwards of 200 hectares. That's not to say that individuals can't participate in rewilding: 'Working with others, whether it's your neighbours or your local allotment group, is one of the best ways of increasing the cumulative, positive impacts of rewilding,' says the organisation.

Here are a few tips from Rewilding Britain on doing things small-scale.

- Break down barriers. Cut holes in fences for easy movement of species that can't fly.
- Reduce or eliminate the use of chemicals.
- Create natural habitats by way of ponds, bug hotels, green roofs, pollinator-friendly gardens and compost heaps.
- If you can, leave a patch of land alone for at least a year to see what moves in.

It's also worth paying attention to what professional gardeners are doing at large public-facing sites too. Many National Trust properties are planting with conservation, nature and climate in mind. The gardening team at Ham House, a seventeenth-century estate and gardens in the London Borough of Richmond upon Thames, has been bringing a touch of wildness to the historic site – not something you'd expect at such a formal place. The gardeners are keeping the lawns longer, to encourage wildlife to move in, and planting certain varieties of apple tree for their ability to adapt to a changing climate. They're also choosing as many flowering species as they can, to ensure nectar is abundant, and creating homes for fungi, moss, insects and frogs in the form of log piles.

A few more green gardening tips

- Use electric tools instead of petrol-powered ones, and borrow them from neighbours or sharing networks.
- Plant more trees and keep the ones you've got. Trees are invaluable carbon sinks.
- Join the seed-saving movement. By swapping and saving seeds among fellow growers, whether that's your neighbour, allotment group or a more formal network, genetic diversity of plant varieties can be preserved and food security safeguarded. The RHS has guides on how to get started and you can find local seed-saving groups through the Gaia Foundation's seed sovereignty campaign.

Nature stewardship: a brighter future

Take a moment now. A breath. A pause.

Close your eyes and think about your happiest childhood holiday. What are you doing? I bet you're at the seaside or a lake; maybe you're camping or rock-pooling. It's warm, you can hear laughter and you've just eaten an ice cream. The chances are that your best memories happened in the great outdoors.

I have vivid memories of family holidays spent at a place called Cultus Lake in British Columbia, Canada. It was a charming community with a very swimmable lake and a big dock to jump off. Cute holiday cottages were within walking distance of the sandy beach and there was a big waterslide attraction nearby. Summer bounties also feature strongly in my childhood memory bank: succulent peaches and cherries devoured in the back seat of our car, bought from roadside stands as my family and I road-tripped through the interior of the province. You haven't lived until you've eaten a baseball-sized, BC-grown peach.

In thinking about how to be the best stewards of nature we can be, the most helpful thing may not be to recommend this habit or that one, but to imagine what kind of experience we want our children to have on this planet. The outdoor pleasures I hope you were fortunate enough to have experienced as a child perhaps might drive you to put into action your own version of what it means to take care of nature. If you live for surfing or sailing, perhaps joining an annual beach clean is a good fit for you. If you're a keen gardener, it could be as simple as teaching your children, grandchildren, nieces or nephews to love rather than loathe bugs. Or, if you were mostly city-bound as a child, perhaps you could join your local community association to improve urban playgrounds where you live.

Whatever nature-orientated activity fills your cup, do that thing, and try also to do something that preserves or conserves it – for the sake of a thriving planet, a continued, comfortable existence here, and for the future memories. As my colleague, the journalist and editor Daisy Greenwell, once said, 'Concerted human effort, for a cause beyond ourselves, is the most beautiful marker of our species.'

And if you can manage to enact that concerted effort, in a calm, consistent way, without feeling too stressed or anxious about it, then that's all the more beautiful.

Links

nationaltrust.org.uk/features/going-peat-free

wildlifetrusts.org/actions/how-go-peat-free

rhs.org.uk/soil-composts-mulches/worm-composting

mains2rains.uk

wildlifetrusts.org/actions/best-plants-bees-and-pollinators

nationaltrust.org.uk/discover/gardening-tips/nine-ways-to-build-a-wildlife-friendly-garden

rhs.org.uk/propagation/seed-collecting-storing

seedsovereignty.info/near-me

Index